CITY, CAPITAL AND WATER

CITY, CAPITAL AND WATER

Edited by Patrick Malone

London and New York

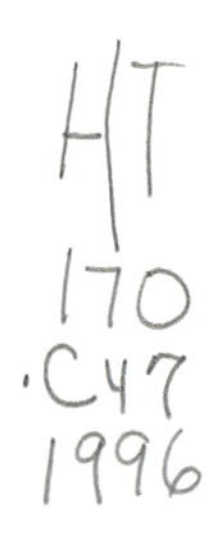

First published 1996
by Routledge
11 New Fetter Lane, London EC4P 4EE

Simultaneously published in the USA and Canada
by Routledge
29 West 35th Street, New York, NY 10001

Typeset in Garamond by
RefineCatch Limited, Bungay, Suffolk
Printed and bound in Great Britain by
Biddles Ltd, Guildford and King's Lynn

British Library Cataloguing in Publication Data

A catalogue record for this book is available from the British Library

Library of Congress Cataloging in Publication Data
City, capital and water/edited by Patrick Malone.
p. cm.
Includes bibliographical references and index.
1. Urban renewal–Case studies. 2. Waterfronts–Case studies.
I. Malone, Patrick, 1946–.
HT170.C47 1996
307.76–dc20 95–52400
CIP

ISBN 0–415–09942–0

CONTENTS

FIGURES

TABLES

CONTRIBUTORS

John Barnes worked as a planning officer at the Docklands division of the GLC between 1983 and 1986, and subsequently as a researcher for the Docklands Consultative Committee (DCC) between 1986 and 1993. He is currently working for an aid agency in Beira, Mozambique. While at the DCC he co-authored a number of reports on the impact of the LDDC on Docklands communities.

Arne Bongenaar is a senior consultant in the field of real estate and infrastructure for De Nationale Investeringsbank NV (The Netherlands). Previously, he was a researcher at the Institute of Spatial Organization TNO in Delft. He is a geographer and has published several articles in the field of regional economics. He is specifically interested in urban development, office markets and office location theory.

Roger Bristow is a senior lecturer in the Department of Planning and Landscape at the University of Manchester (England). He has also taught at the University of Hong Kong. He undertakes research in comparative planning and is currently working on the planning systems of Malaysia and Singapore, as well as recent developments in Taiwan, Hong Kong and China. He has taught and worked on Pacific Rim economies and has written about British and European planning systems.

Bob Colenutt worked as a community planner for the North Southwark Community Development Group in London between 1972 and 1979. He then worked for the Joint Docklands Action Group between 1979 and 1984 and was involved in community campaigns and action research projects around the development of Docklands. He was head of the Docklands Team at the GLC between 1984 and 1986, and head of the Docklands Consultative Committee Support Unit between 1986 and 1993. He has been head of Thames Gateway Unit at the London Borough of Barking and Dagenham and is currently at Haringey Council, London. He is the author of a number of reports and articles on planning and urban regeneration.

Maurice Daly was, until recently, McCaughey Professor and head of the Department of Geography, University of Sydney. He was formerly the director of the Planning Research Centre at the University of Sydney, and director of the Research Institute for Asia and the Pacific. He has worked on urban and regional development issues in Canada, the United Kingdom, Nigeria, India and throughout the Pacific Asian region.

Michael Goldrick is a graduate of the London School of Economics and is currently an associate professor of Political Science at York University, Canada, where he specializes in the political economy of cities and labour. He was a full-time alderman on the councils of the City of Toronto and the Government of Metropolitan Toronto in the mid-1970s. Since then, he has maintained involvements with labour organizations and social policy groups.

Ken Greenberg (Principal, Berridge Lewinberg Greenberg Dark Gabor) joined the City of Toronto in 1977 after a number of years in architectural practice. He founded and directed the Division of Architecture and Urban Design of the City of Toronto's Planning and Development Department. This group focused on the design of public spaces in downtown Toronto, particularly in the historic St Lawrence area and the Yonge corridor, and on the redevelopment of Toronto's Railway Lands. In 1987, he left the City of Toronto to form the planning and urban design firm of Berridge Lewinberg Greenberg. The firm has focused on the redevelopment of inner-city areas, on projects such as the intensification of Toronto's Main Streets, the redevelopment of the Frente Portuario in San Juan, Puerto Rico, and the Faubourg Quebec in Montreal. It has also worked on the promotion of dense, mixed-use communities in peripheral areas. He has recently completed a study for the Province of Ontario on 'Shaping Growth in the Greater Toronto Area'. He chairs an International Advisory Committee for the redevelopment of the Amsterdam waterfront and is an adviser to the Waterfront Regeneration Trust in Toronto. In association with Antoine Grumbach, Architect, Berridge Lewinberg Greenberg has recently won a competition for the Secteur du Moulon, a new urban quarter on the outskirts of Paris. Ken Greenberg has lectured widely in North America and Europe and has taught at the universities of Toronto, York (Canada) and Montreal. He has contributed to journals and to books on urban design.

Hugo Hinsley is an architect in London, working on housing, community buildings and urban planning projects. Recent consultancies have been with the Spitalfields Community Development Trust, London; the Kop van Zuid team, Rotterdam; and the Aboriginal Housing Company, Sydney. He teaches in the Housing and Urbanism Programme of the Architectural Association Graduate School in London. He is an associate of the Development

Planning Unit, University College London, and has taught there and at many other universities.

Patrick Malone is director of the masters course in Urban Design and Regeneration at the University of Manchester. He is a graduate of the Architectural Association, London, the University of Sussex (MA) and the University of Dublin, Trinity College (Ph.D.). He has undertaken research in the areas of urban design and development economics, and is currently preparing a textbook on urban design.

Roy Merrens is a professor of Geography at York University, Canada. He did his undergraduate studies at University College, London and completed a Ph.D. programme at the University of Wisconsin. A former Guggenheim Fellow and a Toronto Harbour Commissioner, he is the author of a number of books and articles published in Canada, the United States and the United Kingdom.

Tetsuo Seguchi is an associate professor at the Department of Regional Planning, Toyohashi University of Technology. He holds a degree in engineering from Nagoya University, and both a masters degree and a doctorate in engineering from the University of Tokyo. He has worked with many urban-planning practices as an adviser, and has published on industrial development, environmental management and regional planning. He is author of the book *Industrial Development and Environmental Management*.

Yoshimitsu Shiozaki graduated from Kyoto University in architecture, and has a masters degree and a doctorate from that university. He is an associate professor of regional planning and housing policy at Kobe University. He is on the board of the Japan Housing Council and Japan Environmental Council, and has researched in housing, marine developments and systems of community participation. Following the Great Hanshin Earthquake, he became involved in the reconstruction of Kobe and the involvement of local groups in redevelopment processes. He has published: *Urban Waterfront* (1987); *Coastal Zone and Open Space* (1991); *Rental Housing Policy* (1993); *Community Architecture* (the 1993 Japanese translation).

ACKNOWLEDGEMENTS

The editor is grateful to the co-authors and for their forbearance in a project that took longer than expected to come to fruition.

Special thanks are due to Dr Aisling Maguire, who corrected the manuscripts and provided an excellent editorial service. The editor is also grateful to Sinead Hennessy and to Roy Merrens who helped to structure the work, and to Rong-Chang Jeng, Yea-Huey Chang and John Archer for their support.

The editor is also grateful to a number of colleagues who helped in the exploration of waterfront projects that are not covered directly in the book:

Tridib Banerjee
Brian Edwards
Bruno Gabrielli
Hans Harms
Brian Hoyle
Allan B. Jacobs
Mark de Jong
Pierre Jurban
Max Kelly
Richard Plunz
Barry Shaw
Marcel Smets
Anne Vernez Moudon
Peter Webber

1

INTRODUCTION

Patrick Malone

INTRODUCTION

This book examines the economic and political forces behind a number of waterfront developments and the relationships between those forces and planning and urban design. It covers projects in seven countries and shows how conditions for urban development changed from the 1980s to the 1990s. The book was prompted by a conference that provided an overview of waterfront projects around the world (Cities on Water, Venice, January 1991). In that about thirty waterfront projects were examined at the conference, it is clear that the waterfront is a major factor in the physical and economic development of many cities. The conference touched on a number of issues, including the size of the projects and their impact on parent cities, the difficulties of acquiring development capital, and the problem of securing social and planning gains from commercial property interests. It did not, however, explore the economic and political frameworks for development. There were few direct references to money, profit or power, or to the motives underlying the redevelopment of the world's waterfronts.

The conference exposed a general tendency to see the waterfront as a unique phenomenon rather than as a frontier that has been invaded by common political, property and economic interests. In the late 1980s, much of the literature on waterfront development was concerned with describing changes in city–port relationships, the causes of decline in old ports, and new projects emerging on the waterfront (Hoyle *et al.* 1988; Hall 1991). Harvey (1989b) presented a more incisive view of the forces behind waterfront development. More detailed examinations followed in the 1990s – many inspired by London Docklands (for example Brownill 1990; Fainstein 1994).

There is scope for more research, particularly into the forces that will shape waterfronts in the 1990s. However, as the Cities on Water (1991) conference illustrated, there is little evidence that the forces controlling planning and design on the waterfront are acknowledged by the professionals on the front line of the development process. This is also obvious from the literature covering the physical aspects of development and design (Torre 1989; Breen and Rigby 1994). This neglect of the underlying forces which shape urban development, and of the relationships between these forces and planning and design, created the impetus for this book.

THE SPECIFICITY OF THE WATERFRONT

The nature of the waterfront may encourage the view that it provides a unique realm for urban development. The decline of old ports and allied transport and industrial areas has allowed the public to regain access to an 'urban edge' which is generally endowed with social meaning (Breen and Rigby 1994). The waterfront has yielded tangible social, recreational and environmental benefits, and powerful interests have raised its political and economic status. It has been used to accommodate personal political ambitions and to house new nodes in the global economy. Its symbolic and economic significance has been reinforced where redundant ports – or reclamation – have provided large tracts of development land close to existing urban centres, in some cases enabling major extensions to the core. The waterfront has also been a target for investment in an era which saw the expansion of finance capital and the global economy (Amin and Thrift 1992). As a venue for the joint operations of financial and property interests it has played an important role in the transfer of pent-up capital into the production of new urban space (Merrifield 1983: 1250).

The typical factors which have made the waterfront available for redevelopment are well documented. For example, Hall (1991) attributes the decline of old ports to containerization, new port technologies, changes in the size and nature of ships and in the transport systems that carry cargo inland. The economics of the port, and the competition between ports, have forced the relocation of harbour facilities to deeper water and sites that offer better inland transport links. This process has created two new frontiers for urban development. It has yielded redundant docklands to be utilized by waterfront cities and the development industry. However, it has also required the development of new ports, allied commercial zones, supporting infrastructure and transportation networks. This book, however, is not concerned with the evolution of ports, or with the factors that have exposed the waterfront to redevelopment. Rather, it is concerned with the economic and political factors behind redevelopment and with the frameworks for planning and design.

The waterfront is remarkable in that it can be endowed with economic, political, social and even cultural significance. Nevertheless, neither the factors that have created the opportunities for redevelopment, nor the processes of renewal, fall outside the common frameworks for urban development. In this respect the waterfront is not unique. It is a new (or retrieved) frontier for conventional development processes; albeit that these processes change over time. Both the types of development and the forms of capital that have colonized the waterfront are common to other parts of the urban structure. Whereas port functions might be said to constitute a specialist branch of the urban economy, redundant docklands and waterfront areas are being invaded mainly by conventional office, housing, leisure and industrial uses. Moreover, unlike the port functions they replace, the land uses emerging in redundant port areas may have little or no relationship to water. That relationship may be purely visual, or limited to leisure activities. Given the size of redundant dockland areas, common patterns of development may take over the hinterland of what was a large port – the suburbanized inner reaches of London Docklands are a good example. In this respect, the redevelopment of old port areas alters the bond between city and water. The area associated with water-based functions may shrink as large redundant ports give way to the 'waterfront phenomenon'. Moreover, as cities lose contact with rivers and the sea, the opportunity to remake the relationship between city and water can depend largely on visual links, leisure activities, amenity values and improved public access to the waterfront. Hence the image of the archetypal new waterfront, although distilled mainly from North American examples, places 'festive' waterside activities against a backdrop of high-density commercial development (Harvey 1989b: 93–5).

Given that the waterfront may be invaded by common land uses and conventional development interests, individual projects may reflect changes in the structure of the development industry, such as the balance of commercial and institutional developers, or the relative distribution of local, national and international interests. London's Canary Wharf is an obvious example in that it demonstrated the significance of the international banking system in the 1980s (Fainstein 1994: 212–13; Debenham, Tewson and Chinnocks 1989: 18).

The tendency to stress the uniqueness of the waterfront may be encouraged by a number of factors. It can provide a platform for flagship developments that convey political ambitions or plug a city or nation into the global economy – functions which may be boosted by the growing emphasis on urban marketing (Ashworth and Voogd 1990). Perhaps because of its political and economic significance, the waterfront is seen as the stamping ground for new urban phenomena. It is presented as an arena for international development capital and as a stage for postmodern culture, the 'post-Fordist economy' and allied changes in architecture and urban design (Harvey 1989a, 1989b). However, it is debatable whether this implies that the waterfront is a unique place where new things are happening, or just a place where the forces of capitalism are currently exercised under a new guise.

The waterfront may mirror economic change as developers attempt to draw rents from lucrative or expanding sectors of the economy. Whether in terms of development or occupying capitals there is a sense that the waterfront absorbs rising players in national and international economic frameworks (although this may have been more evident in the 1980s than in the ensuing period of recession). Before it slid into bankruptcy, London's Canary Wharf symbolized the city of finance capital and the 'yuppie' culture that fed on the expansion of the financial sector and the global economy (Whimster 1992). Later, in London and elsewhere, the waterfronts that provided a stage for the rising economies of the 1980s became symbols of recession and stasis.

It is important to note, however, that no two developments examined in this book share exactly the same economic and political origins. Economically, they may have common roots in monetarism, deregulation, and the conditions for finance and development capital. In terms of the development process, the different projects may demonstrate the marginalization of planning and the state-sponsorship of private development interests. They may also be examined as forms of exploitation of new economic markets and of the political opportunities presented by urban development. In short, they demonstrate factors that have formed part of the backdrop to the 'waterfront era'.

The office space that spreads over redundant docklands provides access to the financial sector and other economic growth areas. Retail, tourism and leisure developments feed on disposable income. Private housing projects are aimed at new markets. However, housing can also be a defining element in waterfront developments, providing mixed-use or inhabited developments that boost inner-city populations and offer social gains, possibly in commercial developments. The emphasis on public open space and facilities is another key factor that may be used to distinguish between developments. Different forces have shaped the redevelopment of the waterfronts of Genoa, Sydney and Barcelona, but these waterfronts fall into the same general category in that they provide public space, access to the water and recreational amenities. The ambitions behind such projects differ from those behind projects driven primarily by economic ambitions or the commercial objectives of the development industry. They may also have a different meaning in terms of the city. Economically, they may be sustained by urban tourism rather than the financial or other sectors of the economy; but they can contribute indirectly to the economic status of the city as 'cultural capital' or resources for urban marketing. They may be imbued with greater social and cultural significance and, as 'public projects', may accommodate a variety of political ambitions. In this respect, the politics of public facilities, exhibitions and festivals have added an important dimension to waterfront development. The provision of amenities may be fused with the appeasement or seduction of an electorate. Moreover, as in Sydney, the waterfront may be used as a 'shop-front' in which the city or nation displays political and cultural symbols for inspection by the rest of world.

As Japanese developments illustrate, port facilities can also be an important element of waterfront and larger marine projects. In addition, Hong Kong and Japan have used artificial islands to create major new airports. But when examining Japanese projects, it is important to distinguish between developments on redundant or existing land and the creation of large marine projects on reclaimed land or on artificial islands created from waste or other material.

Two conclusions can be drawn from this brief review. First, that the waterfront is not a unique realm of urban development, but rather a frontier on which common processes have taken a contemporary form. Second, that although waterfront projects vary in physical, economic and political terms, they may share a common background in terms of economic deregulation, ambitions for flagship projects, or rivalry in the global economy. The form of any individual development reflects an underlying mix of economic and political intentions and the conditions for planning and development. Politically orientated projects that provide large-scale public facilities differ from projects based in economic objectives. In addition, the context for development varies in line with attitudes to planning and social goals, urban design, conservation and urban ecology. Basic factors, such as the size of projects or whether they involve new or existing land, determine physical and land-use structures and the emphasis placed on infrastructure, transport, and physical and economic relationships to the host city and property markets.

It is possible to conjure images of the archetypal 'new waterfront': prestigious office space, financial capital, the global economy, urban tourism, new leisure activities and new models for inner-city housing. The language of this imagery may stretch from high-tech to the hybrid historicism of postmodern architecture; from the 'festive people-place' clichés of urban design to the Soviet-like 'quantitative planning' of large Japanese artificial island cities. One common image highlights the rent-hungry postmodernism of office developments built for the internationalized financial sector, fronted by the 'exploitative conviviality' of leisure activities based on disposable income (Harvey 1989b: 91–8). Harvey's overall image is one of commodification, political contrivance, and the alliance of aesthetics, rent and profit.

The common images of waterfront development are based largely on a small number of projects, notably London Docklands, Boston, Baltimore and Toronto. However, waterfront developments vary within wide limits set by a range of economic and political forces. They reflect common and individual elements, and provide a number of different opportunities to examine how these forces come together with planning and design.

THE WATERFRONT PROJECTS

It is difficult to untangle the knot of forces behind any waterfront development. Projects can be examined, however, on the basis of their underlying objectives and the power structures that generate and control them. It is appropriate, therefore, that two chapters in this book deal with London Docklands. While the waterfronts of Boston and Baltimore are quoted as physical examples of waterfront regeneration (Breen and Rigby 1994: 109–19), London Docklands might be regarded as a 'model' for waterfront development under a monetarist government.

In Chapter two, Barnes, Colenutt and Malone explore the massive push given to London Docklands by the Thatcher governments. They show that the governments' support may be measured in terms of financial subsidies granted under a regime of state-sponsored deregulation, the powers granted to the London Docklands Development Corporation (LDDC), the reformation of planning, or the marginalization of existing local authorities. As an exercise in privatization, London Docklands symbolized Thatcherism in the 1980s. The significance attributed to Docklands in the mid-1980s reflected the bullish ideology of the Thatcher government and the rise of London as a financial centre (Thornley 1991: 162–84). Docklands and Canary Wharf emerged against a background of heady investment and political 'boosterism'. It demonstrated how, with the backing of key political figures, a development might be launched in a deregulated economic environment using a powerful special development agency.

In Chapter three, Hinsley and Malone examine the results of the LDDC's programme in terms of planning and urban design. They argue that deregulation has not been conducive to good planning and that an 'open-door' policy for development capital failed to produce coherent physical or land-use structures. Development in the first stage of Docklands was particularly disjointed. Planning, seen as an impediment to development and likely to frustrate the flow of capital, was restructured. Controls were relaxed while the state sponsored development processes. Later, however, it became clear that the suppression of planning in the first stage of Docklands' development led to projects that were substandard in terms of planning and design, and to projects that failed to exploit fully the potential for development profit (Brownill 1990: 146–7, 174; Buchanan 1989: 39). Canary Wharf forced planning and urban design on to the LDDC's agenda as factors that could reduce development risks and boost the likelihood of profits on the scale envisaged by the developers Olympia and York.

Apologists for London Docklands argue that the LDDC's initial approach was necessary in order to stimulate development interest (Attwood 1989: 122). However, it can be interpreted as a project that marginalized planning and favoured capital only to produce development that fell short of its potential. Ultimately, Docklands came to symbolize recession, the failure of property markets and of an overtly monetarist ideology. It also symbolized the end of Olympia and York, and marked a major event in the history of the property industry.

Whatever the mistakes of the LDDC, both chapters on London Docklands cast doubts on the role of planning and design, and the future for planning in the 1990s. Given that responsibility for London Docklands does not rest entirely with the LDDC and the government, this raises the question of whether the professions were capable of delivering successful strategies for Docklands, or failed to push through satisfactory alternative strategies for development. While planning was marginalized in the first phase of Docklands, the LDDC and development interests eventually adopted urban design in order to generate frameworks for development. In this respect, Docklands might be taken to illustrate the limitations of form-based approaches to design and weaknesses in the design professions.

The marginalization of planning and urban design is also the central topic of Chapter four, which covers the Custom House Docks project in Dublin. Malone examines the extent to which this development has been formed by the ambitions of political and economic interests as opposed to the objectives of planners or designers. He argues that the development process can be institutionalized in planning and legislative frameworks, but that the origins of any project are more likely to lie in the motives of the interests that instigate and control the development process. However, the negotiations between these interests are generally secret and inaccessible to research.

There is evidence that planning processes were hijacked by the interests behind the Custom House Docks, and that, overshadowed by the marketing process, designers were employed to provide imagery to launch the project. Thus, at least part of the energies of planners and designers was used in marketing, rather than for dealing with the issues and problems facing the development. This raises issues regarding the role of professionals (and particularly architects) in the development process, and the capacity and freedom awarded to professionals to operate from a position of 'best practice'.

The Custom House Docks project also demonstrates that a remarkable contrast can exist between the imagery used to launch a development and the reality of the built project. This gap reinforces the view that architecture is used to secure acceptance of projects by the public, to lure investment capital and promote demand. Architectural and marketing imagery may also mask the real nature of the political and economic intentions behind a project. However, if the marketing process means that professionals obscure, or fail to address, planning and design problems that may frustrate development and carry economic costs, then development interests may fall victim to their own marketing strategies and to the willingness of professionals to provide masking imagery.

In the 1980s, urban marketing and the flow of international architectural capital contributed to a period that saw the production of architectural 'confectionery' or images that were intended to please or attract. The redevelopment of the waterfront in Dublin, for example, was fronted by the 'people-place' language of earlier waterfront developments in the United States – imagery that has been used to launch other waterfront projects around the world (see, for example, Process Architecture 1990). This imagery also influenced the redevelopment of Sydney's Darling Harbour, which is the subject of Chapter five.

Daly and Malone trace the roots of the Darling Harbour project to the economic and political structures of Australia in the 1980s. This project grew from a number of factors: the political opportunism of Premier Neville Wran; Australia's wish to enlarge its role in the global economy; and Sydney's bid to become a world city. It also reflected Australia's desire to draw income from foreign tourism. The project originated from strong political motives and has been criticized in terms of the forces behind it. Some of the factors associated with the development have drawn strong adverse public reaction, especially a monorail that connects the project to the central area of Sydney. But Darling Harbour shows that a development can be assessed either on the basis of its origins or on the values attributed to a finished project by the public – particularly where a development provides open space, access to the waterfront and public facilities. Whatever the doubts surrounding its origins, Darling Harbour is relatively successful in terms of its contribution to tourism and as a public project on Sydney's waterfront. In short, Sydney shows that relatively successful or popular projects may spring from dubious intentions.

In Chapter six, Bristow examines the rationale of waterfront development in Hong Kong. He argues that the waterfront presents particular locational advantages, but accommodates the common processes of urban development that are based on property rights and the ambitions of property and political interests. He uses the case of Hong Kong and its new airport complex to show how general economic and political forces can be focused on a specific site.

Two things make this project remarkable: its size and its political context. In addition to the airport (which covers 1,248 ha), it includes a new railway, connecting motorways, tunnels and bridges, two harbour areas and the first stage of the Tun Chung new town. The relationships between Hong Kong, China and Britain have been played out over the project. Moreover, it has been influenced by factors specific to Hong Kong, for example the characteristically close relationship between Hong Kong's political and development interests. Bristow traces the elements particular to the project. He argues, however, that beneath these factors the development is moulded by the general logistics of urban development in capitalist economies.

As in Hong Kong, the large Japanese waterfront and marine developments of the 1980s were strongly supported by national and local governments. Chapters seven and eight examine the great emphasis laid on waterfront projects in Japan, which has led to the reclamation of large areas of land and the creation of artificial island cities. Japan's major waterside cities are expanding by means of marine developments initiated by city governments under the umbrella of metropolitan planning policies. This approach to the waterfront, however, has passed through a series of changes, moving from port-orientated projects, which fed the industrial base of the economy, to mixed-use 'urban' projects in the period of economic expansion and deregulation in the 1980s. Japan's large marine developments now face a period of recession, financial crisis and depressed property markets.

In general, the relatively strong metropolitan governments created Japanese waterfront and marine developments within fairly rigid planning frameworks. But this is not to say that they pursue the best planning objectives. As Japanese waterfront projects demonstrate, 'quantitative planning' can be a ruthless servant of crude capitalism.

To illustrate this point, Shiozaki and Malone examine three island cities in Tokyo, Osaka and Kobe. These projects are regarded as relatively unsatisfactory in terms of housing, transportation, retail, health and leisure facilities. They are also weak in terms of wider environmental values and planning criteria relating to their host metropolitan areas. Thus, economic pragmatism takes precedence over social and planning values, notwithstanding the use of planning tools to order development projects.

This issue is taken up in Chapter eight (Seguchi and Malone), which offers a detailed examination of a large marine development in Tokyo Bay. The Tokyo Waterfront Subcentre project covers 433 ha, and is both an island city and a physical extension of Tokyo and its urban economy. The project is also linked to the development of central Tokyo in that development rights in the core can be traded by private interests for access to development contracts in the Subcentre project. The metropolitan government is promoting this policy as part of the solution to Tokyo's urban problems. The Subcentre project and expansion into Tokyo Bay are identified as factors that can relieve problems arising from congestion: housing shortages; traffic congestion; the under-provision of public open space; pollution and environmental issues. But critics doubt whether the Subcentre project's potential impact on Tokyo and on the city's problems is positive. They suggest that it is led, not by urban and regional planning objectives, but by powerful political and economic ambitions. They also point to the Japanese tendency to marginalize public opinion and criticism of official planning policies. These issues highlight the nature of the public and private interests behind Japan's pragmatic approach to urban development and they show how social values can be marginalized in a system that emphasizes the quantitative rather than the qualitative nature of planning.

In contrast, the planning system in Toronto, as examined by Greenberg and later by Goldrick and Merrens (Chapters nine and ten), points to an 'active struggle' between economic and social objectives, and highlights Canadian efforts to marry planning goals and the ambitions of property and development capital. While it is argued that Canada has sought a middle way between public and private objectives (Ashworth 1990: 136), Greenberg puts this view to the test by examining the intentions behind seven waterfront projects in Toronto. He shows how the approach to the redevelopment of the waterfront has, over time, reflected ambitions for better planning and architecture. There has been a move away from large commercial megastructures and a return to the traditional street and urban block, and values inherent in the urban spatial structure. Free-standing projects that turned their backs on the city and segregated urban functions and circulation have gradually given way to conservation, mixed patterns of land use, traditional block housing and traditional forms of public space. The emphasis on the urban social realm has raised questions about appropriate land uses for a public waterfront and the need for greater public access to areas previously dominated by port and industrial functions.

Greenberg suggests that architecture, urban design and planning are gradually moving in the right direction. In the shadows of Greenberg's analysis, however, old questions linger regarding the underlying purposes of development, the struggle between economic, political and social goals, and the scope for 'best practice' in planning and design. Ultimately, Greenberg demonstrates that planners and architects can evolve better criteria, but they work within limits imposed by political and economic structures.

This question is taken up by Goldrick and Merrens, who explore the potential for ecological and comprehensive planning in Toronto. Through the agency of a federal Royal Commission (and later a Waterfront Regeneration Trust Agency) an attempt has been made to evolve an ecosystem approach to planning in Toronto's watershed. This evolved partly in response to the public's reaction to the 'ceramic curtain' of commercial projects that emerged on Toronto's waterfront in the 1960s and 1970s, and partly because politicians have acknowledged the potential political costs of exploitative development. However, the emphasis on comprehensive and ecologically based planning has exposed the inadequacies of a weak planning system run by a tangled web of government interests and public agencies. Moreover, environmentalism is a weak force within political and economic structures that are based on expansion, consumption and the status quo. The evolution of Toronto's ecosystem should provide better and more effective planning. But Goldrick and Merrens conclude that, whereas the Royal Commission promised much, the changes in Toronto's approach to the waterfront have been confined to piecemeal and administrative reforms. They find that the core processes of urban development in Toronto have yet to be reformed.

The issue of social planning is also discussed by Bongenaar and Malone in an examination of development proposals for the waterfront in Amsterdam (Chapter eleven). In 1985, a large area of redundant dockland along the River IJ was divided between two development projects. The eastern port area was designated for a new residential district of roughly 8,000 dwellings. The IJ-oevers area, which abuts the city centre, was earmarked for a project based on office development. However, while the municipality proceeded with the eastern port area, the IJ-oevers project has not yet materialized.

The failure to develop the IJ-oevers area is due to a number of factors. The proposed project was a late starter in terms of development cycles and the recession. In addition, the area has to compete with other office locations in Amsterdam and the Randstad, and with a tendency towards suburbanization in the pattern of office location. Furthermore, unlike Europe's major financial centres, Amsterdam attracts little demand from the international office market.

In some respects, the IJ-oevers project exemplifies the situation for waterfront projects that failed to get off the ground. It might be compared, for example, with projects proposed for the waterfront in Antwerp (Vanreusel 1990). However, the IJ-oevers project demonstrates the particular nature of Dutch planning. Unlike other waterfront developments (in Dublin or London, for example) the project was not propelled by political patronage, or propped up by state-sponsored financial incentives. Neither was it allowed a relaxed planning regime. In this respect, the project proposed for the IJ-oevers exposes the 'ethics' of Dutch planning in terms of social values and the sensitivity of politicians to public reaction to large development proposals in Amsterdam.

The IJ-oevers project is also interesting because it is being restructured to fit local opportunities and the constraints of the city's office market. As such, the project may hold lessons for other faltering developments centred on the provision of office space.

THE LARGER VIEW

This book examines just some of a large number of recent waterfront developments. Given the scale of the waterfront phenomenon, certain major projects are not covered; for example, developments in cities such as Vancouver, Rotterdam, Boston, Manhattan, Yokohama and San Juan. Many of these projects deserve attention, some because of their size, others because they represent relatively successful attempts to devise new approaches to urban planning or to the waterfront. Rotterdam's Waterstad and Kop Van Zuid projects are obvious examples, if only because the municipality has invested heavily in projects that are shaped partly by social objectives; at least within the terms set by many other waterfront developments (Pinter and Rosing 1988: 114–27; Hall 1991: 20). Other waterfront projects that have incorporated a strong emphasis on housing are interesting as they may prove to be more recession-proof than commercially based developments, and may spearhead the evolution of physical, land-use and transport policies for the city of the 1990s.

It is also important to stress that the waterfront has given rise to a great range of projects and forms of development. Developments vary greatly in nature and size. The 'waterfront industry' in the United States, as represented by the journal *Waterfront World* and The Waterfront Centre in Washington, DC, stretches from modest recreational and fishing developments in small towns to large urban projects involving major international development interests. Arguably, the 'industry' has developed to a point where it is possible to identify different forms of specialist capital operating on the waterfront; for example, in the development of marinas.

While waterfront development has taken many forms, the era of the waterfront has been unstable and dynamic. In the mid-1980s, the availability of large areas of centrally located development land combined with a surge of investment activity to create a high point at which city authorities, politicians and property interests entertained visions of large-scale development. Development activity was stimulated by economic and political factors, liberalized planning frameworks and the state-sponsorship of development processes. However, many cities are now faced with the question of how to adapt aborted or incomplete projects to the circumstances of the 1990s, to recession and declining levels of demand. It could be argued that some projects ultimately will be pushed across the vague line that separates success from failure. Some projects, such as those in Sydney, Barcelona, Rotterdam and Genoa, may be considered successful while others continue to draw a great deal of criticism. Clearly, however, waterfront developments offer an important opportunity to examine the structures that underpin urban development, and so provide lessons for the waterfront and the city.

BIBLIOGRAPHY

Ambrose, P. (1986) *Whatever Happened to Planning?*, London: Methuen.

—— (1994) *Urban Process and Power*, London: Routledge.

Amin, A. and Thrift, N. (1992) 'Neo-Marshallian nodes in global networks', *International Journal of Urban and Regional Research* 16: 571–87.

Ashworth, G. J. and Voogd, H. (1990) *Selling the City: Marketing Approaches in Public Sector Urban Planning*, London: Belhaven.

Atkinson, R. and Moon, G. (1994) *Urban Identity in Britain: The City, the State and the Market*, London: Macmillan.

Attwood, C. (1989) 'London Docklands: urban design in regeneration', in A. Cortesi, *The City Tomorrow*, Florence: Centro Internazionale di Studi sul Disegno Urbano.

Breen, A. and Rigby, D. (eds) (1985) *Urban Waterfronts '84: Towards New Horizons*, Washington, DC: The Waterfront Press.

Breen, A. and Rigby, D. (1994) *Waterfronts: Cities Reclaim their Edge*, New York: McGraw-Hill.

Brownill, S. (1990) *Developing London's Docklands: Another Great Planning Disaster*, London: Paul Chapman.

Buchanan, P. (1989) 'Quays to design', *The Architectural Review* 1106: 39–44.

Budd, L. and Whimster, S. (eds) (1992) *Global Finance and Urban Living: A Study of Metropolitan Change*, London: Routledge.

Clout, H. (ed.) (1994) *Europe's Cities in the Late Twentieth Century*, Utrecht: The Royal Dutch Geographical Society and University of Amsterdam.

Cullingworth, B. J. and Nadin, V. (1994) *Town and Country Planning in Britain*, London: Routledge.

Daniels, P. W. and Bobe, J. M. (1993) 'Extending the boundary of the City of London? The development of Canary Wharf', *Environment and Planning A* 25: 539–52.

Deakin, N. and Edwards, J. (1993) *The Enterprise Culture and the Inner City*, London: Routledge.

Debenham, Tewson and Chinnocks (1989) *Money into Property*, London: Debenham Tewson Research.

Fainstein, S. S. (1991) 'Promoting economic development: urban planning in the United States and Great Britain', *Journal of the American Planning Association* 57(1): 22–33.

—— (1994) *The City Builders*, Oxford: Basil Blackwell.

Frankel, E. G. (1992) 'Artificial islands city developments', *Aquapolis* 5/92: 20–5.

Hall, P. (1991) *Waterfronts: A New Urban Frontier*, Berkeley: University of California.

—— (1992) 'Learning lessons from Docklands', *Planning in London* 3 (September): 10–11.

Harvey, D. (1989a) 'Downtowns', *Marxism Today* 33(1): 21.
—— (1989b) *The Condition of Postmodernity*, London: Blackwell.
Hoyle, B. S., Pinder, D. A. and Husain, M. S. (eds) (1988) *Revitalising the Waterfront*, London: Belhaven.
International Centre, Cities on Water (1991) *Waterfronts: A New Urban Frontier*, Venice: Cities on Water.
Jessop, B. and Stones, R. (1992) 'Old city and new times: economic and political aspects of deregulation', L. Budd and S. Whimster (eds) *Global Finance and Urban Living: A Study of Metropolitan Change*, London: Routledge.
Kearns, G. and Philo, C. (1993) *Selling Places: The City of Cultural Capital, Past and Present*, Oxford: Pergamon.
King, A. D. (1990) *Global Cities: Post-Imperialism and the Internationalization of London*, London: Routledge.
Merrifield, A. (1983) 'The Canary Wharf debacle', *Environment and Planning A* 25: 1247–65.
Pinter, D. and Rosing, K. E. (1988) 'Public policy and planning of the Rotterdam waterfront: a tale of two cities', in B. S. Hoyle, D. A. Pinder and M. S. Husain (eds) *Revitalizing the Waterfront: International Dimensions of Dockland Redevelopment*, London: Belhaven.
Process Architecture 89 (1990) *Benjamin Thompson and Associates*, Tokyo: Process Architecture Publishing Company.
Smith, M. P. (1988) *City, State and Market: The Political Economy of Urban Society*, Oxford: Blackwell.
Thornley, A. (1991) *Urban Planning Under Thatcherism: The Challenge of the Market*, London: Routledge.
Torre, A. L. (1989) *Waterfront Development*, New York: Van Nostrand Reinhold.
Vanreusel, J. (ed.) (1990) *Antwerp: Reshaping a City*, Antwerp: Blonde Artprinting International/City and River Project.
Whimster, S. (1992) 'Yuppies: a keyword of the 1980s', in L. Budd and S. Whimster (eds) *Global Finance and Urban Living: A Study of Metropolitan Change*, London: Routledge.
White, K. N. *et al.* (1993) *Urban Waterside Regeneration: Problems and Prospects*, Chichester: Ellis Horwood.
Zukin, S. (1992) 'The city as a landscape of power: London and New York as global financial capitals', in L. Budd and S. Whimster (eds) *Global Finance and Urban Living: A Study of Metropolitan Change*, London: Routledge.

2

LONDON: DOCKLANDS AND THE STATE

John Barnes, Bob Colenutt and Patrick Malone

INTRODUCTION

Since 1970, every government has placed the redevelopment of the London docklands high on the political agenda and at the forefront of the development of British urban policy. During the 1980s a new era was initiated, as the area now known as Docklands absorbed massive investments in development capital and seemed set to become a fashionable residential and business address.[1] The redevelopment of the area, which covers over eight square miles (20 km^2) and contains 55 miles (88 km) of waterfront, gave the term 'docklands' a new meaning (Fig. 2.1). It suggested a world created by the enterprise culture, the era of the 'yuppie' and the eastward extension of London's financial core. Luxury waterfront apartments and postmodern office buildings became the physical symbols of this new era. For a time the momentum seemed unstoppable. It was as if Thatcherism had triumphed. But in the late 1980s there was an abrupt change. The property market crashed and the credit boom came to an end. In 1992, the giant Canary Wharf office development (Fig. 2.2), the most significant symbol of dockland regeneration, lurched into financial crisis. Docklands became a speculator's nightmare and the decline of Canary Wharf into bankruptcy marked the end of the Thatcherite dream.

In the 1980s Docklands was a paradigm for waterfront redevelopment all over the world. By 1990, it symbolized the plight of a development system that marginalized planning and social concerns and championed monetarism as the driving force in urban development. This chapter presents an analysis of the factors that have made the redevelopment of the London docklands so critical for government. It examines why government intervened, the nature of its interventions and the general lessons for urban policy.

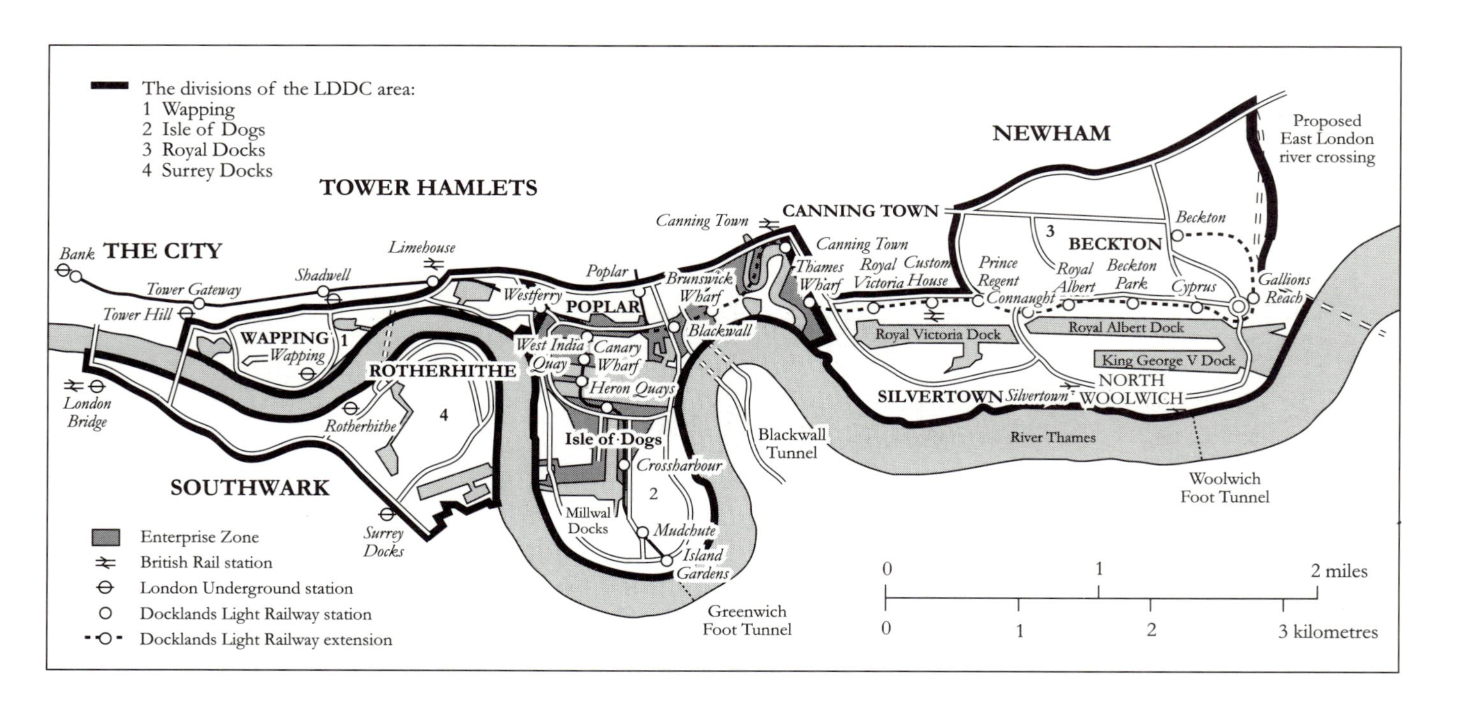

Figure 2.1 London Docklands: map of LDDC area
Source: Philip Ogden

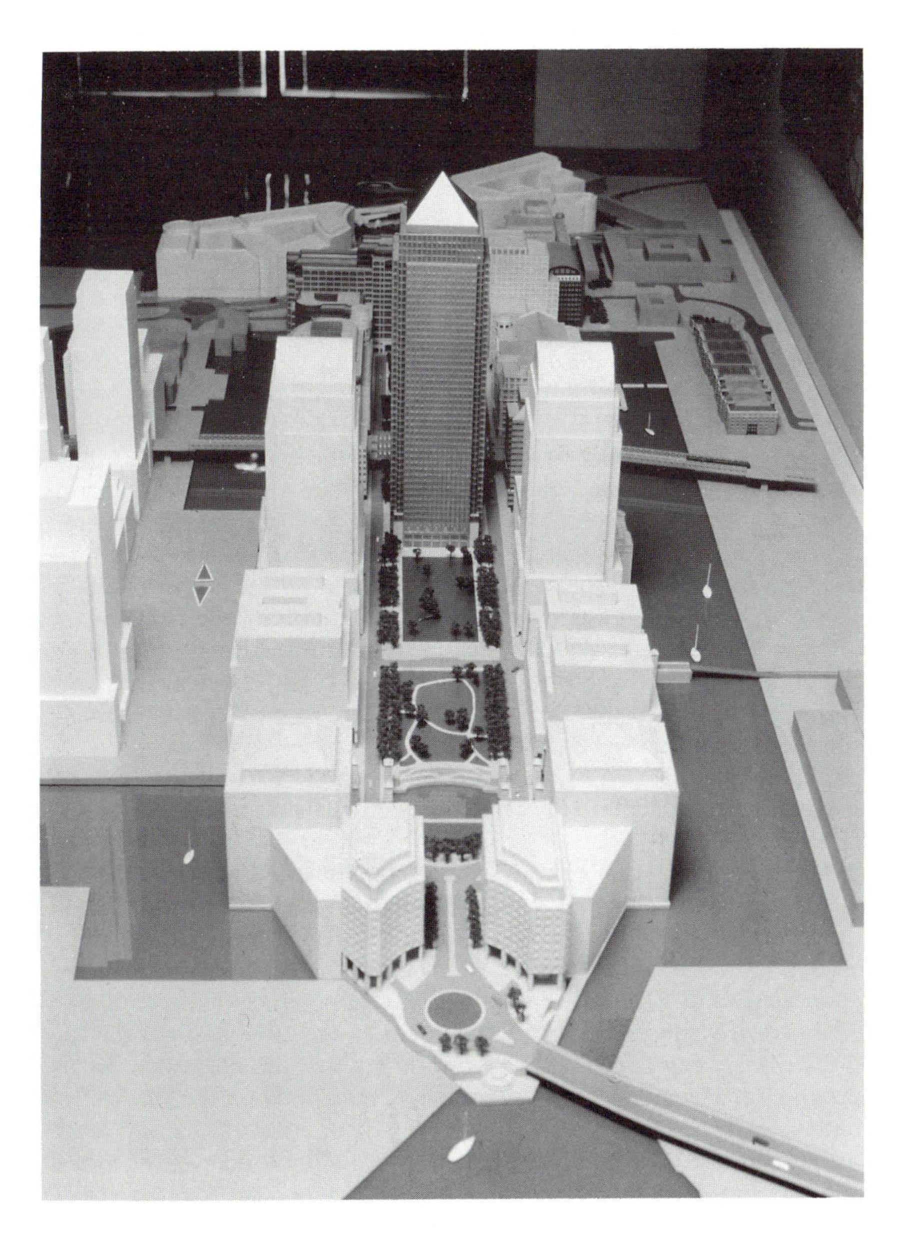

Figure 2.2 London Docklands: model of Canary Wharf
Source: The Builder Group

J. BARNES, R. COLENUTT AND P. MALONE

THE STATE AND DOCKLANDS

Over the past three decades, East London has undergone a dramatic social and physical transformation. In the 1950s the East End accommodated a thriving port and an important industrial base. By the mid-1970s, the area was in a state of serious economic decline, marked by rising unemployment and large-scale dereliction. To some outsiders the area conveyed an image of troublesome dock workers, strikes and social unrest in a remote part of London. To more sympathetic observers, the East End and the port were associated with the Labour movement, with a history of industrial struggles and with the grinding poverty and hardship endured by a succession of immigrant communities. The history of state intervention in the docklands in the period since 1970 reflects these two opposing views.

That wider history of the state's involvement with the docklands begins in the nineteenth century and may be described in terms of a succession of major agencies: the Port of London Authority, the dockland local authorities and, more recently, the London Docklands Development Corporation. The first significant intervention by the state came at the turn of the century, when the confused nature of wharfage and dockland companies threatened the efficient operation of the port. A Royal Commission of 1902 recommended that a single authority should take responsibility for the administration and regulation of dock labour and the ownership of hundreds of acres of land and dock facilities. The Port of London Authority (PLA) was constituted in 1908, when it initiated a period of comprehensive modernization and expansion in the port.

The port and the East End of London required extensive post-war reconstruction in the 1950s. However, it was only when the first dock closures of the late 1960s released large areas of vacant PLA land that central government began to take a strategic interest in the economic future of the London docklands. This interest was not prompted by the decline of the docks, or the shrinking industrial base in the East End, but by the extent of the land that was becoming available for redevelopment close to the centre of London and its financial core. But the full importance of this land was not immediately apparent. When the docks began to close down in the late 1960s, property and financial capital had little interest in docklands. Moreover, the expansion of the financial core and a boom in the office property market in the early 1970s were accommodated without much incursion into the area. Speculative interest in the docklands was restricted initially to fringe sites close to the City of London, but these developments (notably at St Katharine's Dock) signalled that overspill from the commercial core might ultimately have a dramatic effect on the future of East London.

Throughout the 1970s and 1980s, the government's commitment to the regeneration of the docklands increased. In the periods of Conservative rule (1970–74 and from 1979 onwards) the government's interest spiralled in proportion to escalating land values and rising expectations of development profit. In the 1980s, speculation was fuelled by the proximity of the docklands to the City and a growing interest in the potential of an eastward push for property and finance capital. The City, as a centre for banking and financial capital and the seat of a separate local authority (the City Corporation), is endowed with significant political power. It is a vested interest that no government can ignore. Even Labour governments are inclined to provide whatever the City needs to function efficiently and profitably.

Tory interest in the docklands was also sharpened by political considerations. The history of the trade-union movement and socialist political organizations associated with the port imbued the area with an ideological significance and presented a political challenge that was felt most keenly by Conservative governments. Within the East End, the strength of organized labour and the Labour Party was rooted initially in the concentration of a large labour force, organized and motivated by the appalling conditions of casual dock work that prevailed until the creation of the Dock Labour Scheme in 1947. The economic importance of this concentration of labour power was evident in 1889, when striking dockers marched out of the docks and into the City. Dock labour continued to represent a significant political force until the demise of the docklands. As late as 1972 the Conservative government viewed the power of the dockers as a danger to British capitalism.

For the Conservatives, the decline of the port represented the elimination of a major forum for organized labour and the removal of undesirable factions which had played a key role in the politics of the docklands. The Port of London Authority had a long-standing ambition to close the West India and Royal docks, regarded as the power base of the trade unions. Moreover, the government and the PLA sought to abolish the Dock Labour Scheme – which was finally accomplished in 1986, seven years after the Conservatives were returned to power.

For the incoming Tory government in 1979, the ambition to reduce the influence of the Labour-controlled local authorities in East London formed a less explicit part of the political agenda that led ultimately to the establishment of the London Docklands Development Corporation. These authorities were seen by the Conservatives and the City as pro-union and anti-business. They were thought (perhaps wrongly) not to favour property and development interests and to be reluctant to grant planning approval for speculative office development on the fringes of the City. As interest in the expansion and modernization of the City increased, the removal of planning controls from the dockland local authorities became an important political and legislative goal for the Conservatives and a major element of their policy for Docklands. In July 1981, the dockland local authorities were marginalized when Michael Heseltine, as Secretary of State for the

Environment, established the London Docklands Development Corporation (LDDC).

The LDDC epitomized the extension of the government's control over planning and its ambition to undermine local government. The introduction of centrally controlled planning mechanisms, such as the urban development corporations and the Enterprise Zones, were part of a broad attack by the 'new right' on the foundations of the welfare state and on Labour's elusive social-democratic consensus. The wider political agenda of the Conservatives, to stimulate economic growth by 'lifting the burden' from private enterprise, required the suppression of labour power, local politics and the socialist ideology.

Although urban policy in the 1980s was characterized by a shift in favour of the market and private interests, it would be simplistic to suggest that power was simply devolved to capital. The government actively suppressed local community and political interests and created new agencies that implemented centralized control over urban development in the interests of capital (Duncan and Goodwin 1988). In essence, the establishment of the free market in urban development involved the state in the sponsorship of private interests and required a range of state interventions and agencies aimed at liberating the capitalist development process. In this, the LDDC played an important ideological role in legitimizing the state's position as a patron of free enterprise in urban development. However, the LDDC and the flow of exploitative capital into the docklands also fostered opposition and created an arena for political confrontation. Docklands became a byword for conflict in the restructuring of urban space: conflict between developers and local communities; between the centre and the locality; between the poor and the wealthy, the powerful and the relatively powerless. Initially, it epitomized a development process that marginalized planning to serve private objectives. Later it became symbolic of an approach to urban development that carries high costs for the state and for capital.

PLANNING AND THE STATE IN DOCKLANDS

In the period since 1970, Labour and Conservative governments have accepted the need for a planning framework for the London docklands. However, the approach taken by Labour in its relatively short period of office (1974–79) may be distinguished from that adopted by the Conservatives, whether prior to 1974 or after 1979. Under Labour, the docklands were defined as a task for planning and as an opportunity to forge a pact between private and public interests – a pact that eluded Labour interests in the mid-1970s. Under the Conservatives, the docklands were first divined as a deep well of development profit in the early 1970s. In the 1980s, the Conservatives returned to translate the docklands into an opportunity for exploitation.

The London Docklands Study

Published in 1973, the London Docklands Study was significant in that it defined the docklands in geographical and planning terms (London Docklands Study Team, 1973). From the outset, however, the issue was whether the docklands, as defined by the Conservatives, had any real meaning in terms of local people and their needs. The Study was primarily intended to highlight development opportunities, to bolster confidence and to create a secure framework for capital investment. It produced five options for development, but the thrust of the Study was characterized as 'bringing the West End into the East End'.

In attempting to impose its agenda on the flow of development capital into East London, the government presented the interests of local authorities in the area as parochial and self-serving. By contrast, development and financial interests were presented as neutral or as being in the national interest. The Study was so insensitive to local interests that it failed to progress beyond the consultation stage. It was finally abandoned, partly in response to pressure from local political and community interests, but also in the face of a recession in property markets, and the emergence, in 1973, of a Labour-controlled Greater London Council (GLC). Nevertheless, the Study highlighted the nature of the underlying conflicts which would come to dominate the redevelopment of Docklands.

The London Docklands Strategic Plan 1976

In January 1974, the London Docklands Study was followed by a new initiative in the form of the Docklands Joint Committee (DJC) which was made up of the GLC, five East End boroughs and co-opted members.[2] The DJC carried Labour policies into the docklands between 1974 and 1981 (when it was ousted by the LDDC). In 1976, two years after it assumed administrative responsibility for the redevelopment of the docklands, it launched the London Docklands Strategic Plan (DJC 1976) as a radical alternative to the earlier London Docklands Study.

The Strategic Plan was, to a large extent, a response to local needs and existing deficiencies. It also reflected contemporary concerns about the role and nature of planning, the growth of community politics and the mobilization of local action groups. In this respect, the commitment of the DJC to genuine social progress cannot be denied. However, its potential for reform was limited by its inability to exert control over private investment and landowners. Moreover, the DJC's Strategic Plan had a fundamental weakness in that, although partly the product of local community and political concerns, it sought to appeal to development capital and investors. As the GLC later conceded, the plan was:

> A compromise between the interests of private developers and the

> local workforce and community of Docklands. On the one hand the plan's overall objective was to secure the redevelopment in the interests of the local community; on the other hand, much of the plan was designed to open up Docklands for private profit by providing infrastructure, in terms of road, public transport and cheap land.
>
> (GLC 1982: 6)

Initially, the ambiguous nature of the Strategic Plan and its balanced approach to investment did provide the basis for a broad consensus, but this could not be sustained. The success of the Strategic Plan was heavily dependent on existing landowners and on their willingness to generate a degree of certainty by providing a framework for investment in infrastructure and transport. It soon emerged, however, that the problems and opportunities which the docklands presented were interpreted differently by different interests. For example, the Port of London Authority (PLA) sought to maximize the development value of its substantial land holdings. In contrast, the British Gas Corporation insisted that its land was required for future operational purposes. Most of the 1,400 acres (566 ha), expected to become available during the Plan's first phase, were eventually released, although much of it required considerable reclamation work. The PLA's growing financial crisis, the costs attached to restructuring the port, and the recessionary state of the London property market, encouraged the PLA to sell land which was bought by the dockland local authorities on the basis that this would enable them to pursue their planning and political objectives. However, the intransigence of landowners continued to raise concerns about the further release of land for the subsequent phases of the Strategic Plan.

The implementation of the Strategic Plan was frustrated by financial constraints imposed by the Labour government in response to the growing economic crisis of the mid-1970s. Cut-backs in state investment and expenditure meant that, regardless of the barrier effect of land ownership, public investment in social and economic infrastructure was restrained. It was estimated that to implement the first phase of the DJC's plan would require £359 million. Less than half that amount was made available. Moreover, as government expenditure on housing in the United Kingdom fell by 55 per cent between 1975 and 1981, the DJC's programme to construct approximately 6,600 dwellings by 1982 fell behind schedule. By March 1981, only 1,500 dwellings had been completed.

Reversing economic decline was crucial to the DJC's strategy, which aimed to recognize the employment needs of the local community and safeguard existing manufacturing industries through land-use zoning. Here, the DJC was constrained by its inability to influence the decisions of industrial and manufacturing interests in terms of their location. Later the LDDC was able to claim that some 8,500 jobs were lost in the docklands in the five years prior to 1982, whereas only 800 new jobs were created against the Strategic Plan's target of 10,000–12,000 new jobs.

The expectations raised by the Strategic Plan may have been dashed, but it was important to the community and to politics in the docklands in that it represented a major concession to local interests. Ultimately, however, its inherent weaknesses, and its failure to deliver gains to the community, further politicized the docklands debate. As the DJC weakened its commitment to the principles of its Strategic Plan, for example by allowing office development and private housing, local reactions and social conflict intensified. Thus the DJC's attempt to manage social conflict through a process of consensus and compromise backfired. Inevitably, different interests pulled in different directions as the limitations of capitalist social relations became clear. In attempting to please all class interests, the Strategic Plan 'was full of ambiguities which left the door open for redevelopment to be carried out according to the outcome of continued class struggle' (Newman and Mayo 1981: 540). This struggle was later drawn into a larger political arena as the incoming Conservative government initiated a broad attack on public expenditure, local government, planning and the Welfare State. Within the docklands, the Conservatives undermined the local authorities by presenting the problems of the area as a crisis of the local state apparatus. Local democracy was deemed inefficient and local interests were characterized as parochial. The DJC was depicted as bureaucratic and ineffective in terms of the aspirations of local people and business. In depicting the docklands as a wasteland where nothing was happening, the government called into question the competence of locally based mechanisms in inner-city regeneration and paved the way for the London Dockland Development Corporation.

The London Docklands Development Corporation

In the early 1980s, it was clear to the Conservatives that the opportunities offered by Docklands could be more efficiently exploited by capital from within an enabling planning structure. It is rumoured that Michael Heseltine was advised against using national legislation to create one urban-development corporation. Together with the LDDC, a second urban-development corporation (the MDC) was created in Liverpool. However, the importance given by the government to the London docklands was evident in the fact that the LDDC was funded at five times the level of Liverpool's MDC. It was also apparent in the importance of Docklands in the evolution of national urban policy in the 1980s, for example in terms of the use of urban development corporations and Enterprise Zones.

In launching the LDDC the government created a new planning framework to secure development profits in the docklands and ensure against risk. The planning powers of the London boroughs of Southwark, Tower Hamlets and Newham were automatically transferred to the government-appointed Board of the LDDC. At the same time, over 600 acres (240 ha) of local-authority land were vested in the LDDC, together with the redundant land of the Port of London Authority in the London Docks and West India Docks.

As in the early 1970s, the ambitions of property and other powerful interests were articulated by central government, and presented as being neutral or in the national interest. The aura of neutrality belied the fact that the policy changes introduced by the Thatcher government, and actively sought by property and financial interests, required extensive legal and fiscal deregulation. These policies also required that greater central control be extended over local government, local expenditure and planning, and that the linkages between the locality and local control be dissolved. Thus the LDDC was established to impose central government policy on 'local space'.

Although the LDDC marked a distinct shift in urban policy towards the interests of capital and the requirements of profit over social needs, it also represented a degree of consistency in British urban policy. State intervention to improve the profitability of private capital is an established feature of British urban policy. Moreover, the LDDC was preceded by a long line of area-based agencies and was partly reminiscent of the British new town corporations. However, it represented a reorientation of the state's involvement with urban development in asserting the primacy of the market and free enterprise and in releasing private development from local jurisdiction. Another important aspect of the state's new apparatus was the appointment of private property interests to positions of power, for example to the boards of the urban development corporations. Nigel Broakes, of the international property company Trafalgar House, was appointed chairman of the LDDC. In 1984, he was succeeded as chairman by Christopher Benson of MEPC, the second largest property company in the United Kingdom.

Whereas Broakes' primary role was to encourage property interests to invest in Docklands, the combined power of the government and the property sector was presented as a major new force that would secure the redevelopment of the area. The LDDC's key objective was to create the appropriate conditions for investment. It pursued this objective in a number of ways. For example, it opted to maintain a fluid situation for development capital by rejecting 'master planning' or an overtly fixed approach to the physical restructuring of the docklands. It circumvented 'traditional' planning procedures and democratically approved local planning mechanisms. The LDDC also removed the burden of social criteria from decision-making processes. In short, it operated a flexible planning regime and a developer-friendly environment intended to encourage investment and property speculation. The needs of developers were given priority over social and community needs. The development process was streamlined and

Table 2.1 London Docklands: land ownership in 1980

Public agencies (acres)		*Local authorities (acres)*	
Port of London Authority	1,317	Greater London Council	154
British Gas Corporation	617	LB Southwark/GLC	300
Thames Water Authority	361	LB Newham	196
British Rail	142	LB Tower Hamlets	91
CEGB	74	LB Greenwich	10
British Waterways Board	22		
Subtotal	2,533		751
Total public agencies and local authorities:		3,284	
Total area of Docklands:		5,500	
Total under water:		528	

Source: Travers Morgan 1973: vol. 1

bureaucratic procedures simplified at the expense of local consultation and accountability. The function of planning was largely subsumed under processes concerned with investment, marketing and image-building. Behind the marketing imagery used to lure development capital into Docklands, the LDDC exploited public funds to prepare land for development, to provide infrastructure and to socialize the costs of private development and the pursuit of development profit.

The planning framework embodied in the LDDC legitimized the government's role in the development process and in the provision of land, infrastructure and the economic and physical contexts for development. Land ownership and the power to redistribute land were central to the LDDC's ambitions for Docklands. Whereas land in the area had been dominated by various public agencies that were not necessarily sensitive to market signals, the LDDC absorbed land into a single agency committed to preparing and releasing that land for private development (Table 2.1).

Partly through compulsory purchase, the LDDC acquired 2,168 acres (800 ha) of land by March 1994. Initially, it established confidence and development momentum by releasing land on favourable terms, although, in the long term, it aimed to establish a trend of rising property values and development profits.

Land ownership lent the LDDC greater control over the type of development that might be undertaken, and over which factions of capital might enter the area. Moreover, it allowed the LDDC to enhance the relative attractiveness of the area for developers by acquiring and preparing land, releasing land at favourable prices and employing public funds to offset acquisition and reclamation costs. Canary Wharf provides a good example of how the LDDC underpinned the development of Docklands with cheap

Table 2.2 London Docklands: share of UK urban development funds (£m)

Year	*LDDC*	*All UDCs*	*Urban total*	*LDDC share (%)*
1981/82	31	38	162	19
1982/83	41	62	283	14
1983/84	62	94	325	19
1984/85	58	88	450	13
1985/86	57	86	436	13
1986/87	65	89	429	15
1987/88	83	160	514	16
1988/89	116	255	598	19
1989/90	256	477	815	31
1990/91	333	607	964	35
1991/92	249	601	979	25
1992/93	156	515	984	16
1993/94	105	371	936	11

Sources: *Financial Times*, 21 February 1991; *Hansard*, 14 February 1991; *Hansard*, 28 October 1994); and data from the Department of the Environment
Note: As data shown for all UDCs include the Docklands Light Railway, total expenditure in Docklands area was higher in some years than urban block expenditure on LDDC suggests.

land. The impact of 10 million ft^2 (950,000 m^2) of commercial space on the credibility and the momentum of Docklands was secured, at least in part, by the LDDC's decision to sell the site for Canary Wharf for £400,000 per acre when surrounding land values were well above £1 million per acre (DCC 1992).

In general terms, the LDDC was given substantial powers, financial resources and political support to secure the redevelopment of Docklands. Despite the emphasis on leverage and 'pump-priming', investment in Docklands has relied on substantial injections of state funding. The state's sponsorship has taken the form of direct grants to the LDDC and indirect subsidies, for example to the Enterprise Zone or through the provision of infrastructure.[3] By 1991, total government expenditure in Docklands was estimated to exceed £2.5 billion. This figure rose to approximately £4 billion by 1995.

In the 1980s, the importance attached by the government to the regeneration of Docklands was remarkable. As a major symbol of state intervention in urban development, Docklands took up a large share of the public money earmarked for the urban renewal. By 1990, it was drawing 35 per cent of the total expenditure for English inner cities (Table 2.2).

DOCKLANDS: VISION AND REALITY

Although the LDDC rejected rigid master planning, it did have a vision for Docklands that embodied objectives coined by economic and political interests. For example, in exploiting the docklands it aimed to boost specific sectors of the economy and to restructure the housing market by promoting home ownership. In this respect, Docklands was shaped by external forces that were 'localized' through the agency of the LDDC. These forces were national and international in scale. They reflected a particular economic phase in capital accumulation, and a specific 'episode' in what Harvey (1989) describes as the restless formation and reformation of urban space. In the case of Docklands, urban space, once shaped by financial capital centred in shipping and industry, was restructured by property capital to provide space for new economic functions. In this, the role of the LDDC has been to siphon external economic forces into Docklands by facilitating property capital and the supply of space to specific forms of capital.

While the LDDC provided an arena for national and international economic and political forces, other factors also impinged on the redevelopment of Docklands: factors such as Docklands' relationship westward to the City and eastward to outer London. In 1993, the significance of Docklands was enhanced by the Thames Gateway initiative, a strategic planning scheme to regenerate the Thames estuary and exploit the Channel Tunnel rail link and potential access to European markets.[4] Docklands has also been used to reduce development pressure in West London. The London Planning Advisory Committee (established in 1986 following the abolition of the Greater London Council) characterized West London as a zone of 'restraint' and East London as a zone of 'regeneration'. The east/west debate was not underpinned by a rigorous understanding of the effects of the geographical redistribution of development pressure on the social and economic structure of London. Nevertheless, it served to legitimize the eastward flow of development capital and to protect politically conservative western areas from over-development.

Although Docklands has been shaped by a variety of forces, the LDDC and the government have sought to present it as proof of the capacity of market-led policies to mobilize private enterprise and reverse two decades of economic decline in East London. Clearly, however, doubts may be raised about the success of Docklands, whether in physical, economic or social terms. In physical terms, the LDDC has effectively re-enforced a tradition of uneven spatial development. The absorption of development capital into Docklands reflects the tendency for capital to concentrate in profitable areas where rents and development profits are highest and most secure. In the case of Docklands, the area was 'magnetized' by its proximity to the City and Central London. Within Docklands, the spatial distribution of investment is again uneven, in that it is concentrated in Wapping, Surrey Docks and the Isle of Dogs, where the LDDC has released between 70 to 80 per cent of its land. However, roughly 50 per cent of the LDDC's land bank remains undeveloped, and, for example, the Royal Docks has attracted only 5 per cent of total capital investment in Docklands. The distribution of development capital within Docklands reflects the potential of the different parts of the area to yield development profits. It also hints at the general failure of the monetarists' claim that, ultimately, the wealth created by a few 'trickles down' to the benefit of all.

The economic and political objectives of the LDDC have promoted the revalorization of the docklands rather than concern for social or planning issues. Those objectives fostered, for example, the development of commercial space and 'executive' housing as the 'twin spearheads of a cumulative process of speculatively led, and financially determined development' (Goodwin 1991: 273). Since 1981, over 27 million ft^2 (2.5 m^2) of commercial floor space have been built in Docklands, together with nearly 17,000 residential units.[5] But the huge output of space in Docklands masked the fact that the LDDC's market-led approach was flawed by contradictions. Its 'success' placed it in a two-way relationship with the recession that hit Docklands and property markets in the late 1980s, and that resulted in the LDDC's objectives being hindered by the logic of the market that it sought to accommodate. In short, Docklands became the victim of a recession to which it had contributed. The release of large areas of land so close to Central London exacerbated a general bout of property speculation while contributing to the massive oversupply of space, falling property values and rising vacancy rates. Initially, the potential gains in Docklands were such that developers felt that they had to get involved. Tax-driven development, for example in the Isle of Dogs Enterprise Zone, encouraged the provision of space without due regard for demand. By 1990, Docklands accounted for more than 50 per cent of the development activity in Greater London. But in 1988, the supply of office space in Central London had already exceeded demand and the take-up of space was soon to drop to half its previous level.

A number of factors aggravated the effects of the property crisis in Docklands. The City Corporation and financial institutions, which originally viewed Docklands as an opportunity, had begun to regard it as a threat to the historic position of the City as the financial core.[6] In 1986, the City Corporation radically revised its Draft Local Plan, clearing the way for the development of 20 million ft^2 (1.8 million m^2) of commercial floor space within the City. The situation in Docklands was further overshadowed in that, after six years of unprecedented growth, employment in the commercial and financial sectors declined by 10 per cent between 1988 and 1991. The LDDC's problems were heightened in that development activity, particularly in the Enterprise Zone, was allowed to run ahead of the infrastructure, thus undermining the image of Docklands for investors. By 1990, the effects of the recession were such that vacancy rates for commercial property on the Isle of Dogs approached 50 per cent. As property values fell, Docklands witnessed the demise of some leading property companies and precipitated a major crisis for financial capital. While in 1995 vacancy rates remain static at 40 per cent, the LDDC's failure to recognize the essential logic of the market and the importance of balancing supply with demand leaves open to interpretation the idea that Docklands was market led. Essentially, planning and social concerns were marginalized to serve a largely inefficient exercise in exploitation. The tragedy of Docklands lies not only in its invalidity in planning or social terms, but also in the repercussions for the state, and the private misfortunes of some of Docklands' key players.

In terms of the state, Docklands has absorbed massive injections of public money. The LDDC was launched on the premise that the prudent application of 'pump-priming' expenditure would initiate a regeneration process that would become self-financing. In fact, increasing levels of development activity have effectively absorbed spiralling levels of state investment, notably for land reclamation and transport infrastructure. By 1991, 25 per cent of the total grant to the LDDC (almost £1.4 billion) had been spent on roads and transportation. In contrast, community and social infrastructure accounted for less than 5 per cent of total expenditure.

Clearly, criticisms may be levelled at the economics of Docklands in terms of its meaning for the state and for many property and financial interests. Moreover, it is clear that the state's sponsorship of Docklands cannot be justified in terms of benefits to the local community. Although the local economy has been dramatically restructured, unemployment in the area in 1991 was higher than in 1981 when the LDDC was established (DCC 1991). Similarly, in terms of housing needs the situation in the dockland boroughs has deteriorated. The number of households accepted as homeless increased by 204 per cent between 1981 and 1988. In April 1988, roughly 22,500 people were waiting for a council house in the dockland boroughs, while approximately 85 per cent of new dwellings in Docklands were beyond the means of the majority of local people.

Considerable uncertainty now surrounds the future of Docklands. Developed and undeveloped areas lie side by side. Canary Wharf contrasts with the remnants of the indigenous environment. Luxury apartments on the waterfront contrast with the public housing of the East End. There are marked disparities in the built environment and in the social conditions of the area's population. Doubts raised in the media and elsewhere commonly focus on images of exploitation and political collusion.

In summary, it might be said that Docklands is a costly failure in planning and social terms. This is a common criticism of Docklands. Less common, however, is the criticism that Docklands is essentially the product of crude, careless and inefficient efforts to generate profit. The inefficiency of Docklands as an economic exercise is evidenced by the fact that the LDDC, in creating a cradle for private capital, was confounded by the difficulties and contractions of the market which it attempted to serve. The failures of the monetarist period in British urban development are now symbolized by Canary Wharf. These failures epitomize a system of urban development dominated by crude economic and political interests, which may marginalize social and planning considerations, but which are also insensitive to the laws of efficient exploitation.

DOCKLANDS AND URBAN POLICY

In the wake of a period of unprecedented state intervention in Docklands, it may be useful to examine its general lessons for urban policy. At first glance, the pattern of state intervention in the docklands since 1970 might be said to reflect opposing ends of the political spectrum, with the middle and late 1970s representing Labour strategies, and the early 1970s, the 1980s and beyond representing the strategies of Tory governments. Labour's approach might be portrayed as an attempt to encompass social and locally determined goals. Conservative policies might be seen to be nationally and internationally derived and determined by the demands of capital, the market and the ambitions of property interests. In some respects, however, Labour and Conservative strategies have not been that different. Both employed state intervention and public funds in a bid to restructure 'local space' and re-establish capital accumulation in the docklands.

Whereas the problems of the East End have always been indigenous to the British social and economic structure, political expediency on the part of Labour and Conservative governments dictated that those problems be delineated in geographical and physical terms. Problems rooted in the socio-economic structure were masked by portraying the docklands as an issue for physical planning. Moreover, Labour's approach to the docklands in the period 1974–79, although sympathetic to local authorities and local interests, was not significantly left wing. It came from right-of-centre Labour politicians, whose view of socialism was centred in municipal politics and 'state capitalism'. However responsive Labour's strategy may have been to local community needs and local political action, its implementation necessitated that it should appeal to capital and property interests. The DJC largely failed in its social and housing programmes and the local community held little sway in its decision-making (Brownill 1990).

The strategy of the Conservatives and the LDDC was more 'radical' and explicit than that of Labour, both as an expression of ideology and in its readiness to override local opposition. But the Conservatives' strategy was not wholly the market-orientated exercise in free enterprise and 'non-planning' it purported to be. It required, and later demanded, state intervention on an unprecedented scale. It employed increasing levels of state subsidy and a legislative framework that provided extensive powers to the LDDC, for example, to prepare and deliver land to property capital.

Labour and Conservative strategies for the docklands are also comparable in that both were frustrated by the 'logic' of capital and property markets and by economic factors beyond the reach of political management. An examination of state intervention in the London docklands shows that the state may be able, temporarily, to solve or deflect difficulties and contradictions in the capitalist development process, but it cannot exert adequate control over these processes. In addition, the LDDC has demonstrated that where the profitability of capital is the expressed intention of the state, that objective cannot be guaranteed. Whereas the property recession of the late 1980s resulted from a wide range of causes, the LDDC fell prey to common market forces and, at a general level, thwarted the profitability of property capital by contributing to the recession.

The urban development corporations are being wound up and the government has evolved new mechanisms for urban regeneration. The emphasis is now on partnerships, a 'new consensus' and the re-alignment of the private and public sectors. Partnerships are promoted by the Conservatives as an approach that is aimed at ensuring that all sides benefit from urban regeneration. The City Challenge programme, for example, was heralded as a balanced and common-sense approach to urban renewal. In this, it reflects the softer face of Tory politics in the post-Thatcherite period. However, whereas the notion of partnerships is not new (local authorities have traditionally worked in partnership with the private sector), collusion between the state and property interests remains an integral part of the political–economic structure. What is relatively new is the degree of ideological emphasis placed on the partnership concept and the specific nature of the state's collusion with property interests. Thus the partnership concept is a confusing mixture of restrained collusion and pluralist hype. It reflects the specific opportunities for capital accumulation in a period marked by recession, the retraction from Britain of international property capital and the post-Thatcherite politics of the Citizen's Charter. The partnership concept is the old wolf in new clothing. It represents an adaptation of the relations between government and capital coined to suit the current economic and political climate, even though the factions of capital that partnerships now accommodate are not the same as those that functioned alongside the government in the 1980s. The current beneficiaries of urban policy may be more national or local in origin, and more evidently concerned with development or construction than with investment capital. In this respect, the new alliances between capital and government may offer shelter to some factions of property and construction capital in the current recession and may help to stabilize the economic environment for specific interests. Moreover, partnerships may reflect the realization that planning, particularly in relation to transport and infrastructure, is a necessary factor in successful and profitable development – a lesson evident in the strained relations that eventually existed between the LDDC and frustrated private interests in Docklands. As the new instruments of urban renewal, partnerships may also be directed at different and more social forms of development. In that they may be concerned, for example, with the renewal of public housing, they reflect new political objectives and a renewed search for political credibility.

The partnership concept is being promoted, not only among business leaders and property developers, but also by the Labour establishment within local and national government. Labour regards partnership as a sensible and realistic route through which Labour politics may return from the wilderness of the 1980s. The question of whether partnerships represent a truly socialist strategy expresses the general dilemma facing socialism and the British Labour Party in the 1990s.

It might be argued that the Conservatives and Labour now adopt similar urban policies, but for different reasons. The Conservatives are in flight from the failures of Thatcherism as symbolized by Docklands and Canary Wharf. Labour is ambitious for a credible pact with the electorate, the City and property capital. The issues for Labour were exposed in the 1980s as local Labour governments, under siege from Westminster, gradually turned their attention to the needs of capital rather than the community. In Docklands, this meant throwing the weight of local government behind efforts to support Canary Wharf and the LDDC. Community groups rejected this defeatism and demanded a new approach centred on community enterprise and the regeneration of the indigenous economy. However, very few Labour politicians took up these issues, and for some, the demands of community groups were seen as an embarrassment that might inhibit investment in Docklands. Thus, for those to the left of the party, Docklands may already signal the underlying contradictions of Labour's position. Experience suggests that where, for example, local authorities signed accords with the LDDC to secure 'planning gains' in Canary Wharf and the Royal Docks, only minimal gains materialized. Labour local administrations embraced a right-wing agenda, only to find themselves compromised. At a general level, the failures of the LDDC as an agency for urban development, or as a nucleus for negotiation and the mobilization of public opinion, should throw doubt on Labour's urge to find comfort in the partnership concept.

Essentially, Labour's difficulty is to find a socially and ideologically credible position on urban development in a capitalist economy. At this moment, however, there is a relative paucity of radical or alternative positions in British planning. Before the property crash in the late 1980s, the LDDC appeared to be riding the crest of a wave. The sheer power of the market as manifested in Canary Wharf was dumbfounding. Although the crash exposed the shallow foundations of the market-led approach, silence has settled around the question of what might take its place in urban regeneration. The Labour opposition has lost faith in the public sector approach of the 1970s, and now believes much of the right-wing criticism of that approach. However, in allying itself with market-led solutions, Labour has to find a credible strategy in relation to the contradictions and shortcomings of the market. Given that the property-led approach to urban regeneration that evolved under Thatcherism is widely discredited, the issue for Labour is how to respond to the 'new consensus'. In this, there is a danger of being diverted by an examination of the mechanics of planning, and by the potential to launch new agencies or mechanisms (for example, adaptations of the Single Regeneration Budget) that might effect small structural shifts in the social and economic conditions for development. Monetarist urban policies have been traded for new policies that imply consensus but do little to alter the structure of relations between government and capital. Essentially, these new policies will allow property and allied interests to continue to restructure the built environment for their own ends. Theoretically, it is possible to envisage policies that would empower communities and promote social criteria in the development process. However, it is more likely that urban development will remain in the grip of property and allied interests, albeit that development may be set within a framework of decision-making that is fronted by the rhetoric of consensus.

CONCLUSIONS

In the search for alternative approaches to urban development, some lessons may be drawn from Docklands. Clearly, Docklands indicates the need for local, democratic accountability and participation in decision-making processes and the more fundamental need for social criteria in planning and urban development. But planning in Britain has slipped the reins of social responsibility. It has suffered an ideological reversal and shed the connotation of social idealism. Moreover, the status and power of planning in the political arena have been reduced. It is difficult to believe that British planners ever confronted the radical issue of social land ownership, or that they could now engage in radical debates, for example on sustainability or the redistribution of resources.

While planning is confronted by a range of issues, the problem is to provide a penetrating analysis of the economic and political forces which shape planning and the urban structure. If the task is to create a realizable strategy, then that strategy must confront the logic of urban development and the power structures that pursue that logic in urban space. Docklands demonstrates that urban development generates profits for a raft of interrelated property and allied interests and that urban space is shaped, in part, by the economic logic of property development and investment. The invasion and exploitation of the development process by property interests is such that it is difficult to see how the concepts of consensus or 'planning gain' can deliver substantial improvements in urban development. At best, the approaches now on offer in Britain will attempt to make demands on property interests without undermining the essential logic of development economics. The 'land debate' in Britain may be dead, but the circumstances that gave rise to that debate exist today as they did in the mid-1970s. The achievement of a valid planning system still depends on a radical reassessment of land ownership and property rights and the need to challenge the profit function in development processes. Given the existing system of development, it is difficult to see how provision can be made for social and environmental programmes, social housing, employment, and a range of other social objectives. Any factor that threatens to reduce or impede the flow of rent or development profit must swim against the tide. In this respect, it is also fair to ask what community empowerment can mean within a broadly capitalist framework for development. How far can wider economic and social objectives be pressed as an alternative strategy in the face of powerful, and perhaps global, economic forces?

Clearly, there is a need to move planning out of the realm of profit-led, opportunistic and rent-centred urban development. There is a need to adopt different values. But the adoption of a wider set of social values is not on the agenda. The political left, in its search for an 'electable pragmatism', offers little challenge to the market bias in urban development. The lessons of Docklands may be obvious, but the power of capital is likely to remain pervasive.

NOTES

1. The term Docklands, as used in this chapter, refers to the area under the control of the LDDC which was established in the early 1980s.
2. The local authorities were Tower Hamlets, Southwark, Greenwich, Lewisham and Newham.
3. Docklands' Enterprise Zone covered an area of 482 acres (195 ha) of land and water. Financial incentives for commercial and industrial development included: exemption from rates for ten years from 26 April 1982; 100 per cent tax allowance on capital expenditure; and exemption for development land tax. In addition, development was subject to a simplified planning regime. Developments that conformed with published requirements did not require planning approval.
4. Heseltine's 1992 East Thames Corridor initiative was relaunched in 1994 as the 'Thames Gateway'.
5. A total of 52 per cent of residential units have been on LDDC land.
6. When Christopher Benson took over the chairmanship of the LDDC in 1984 he suggested that Docklands should be supportive of the City, and that the City should maintain its unique identity. Benson held that if Docklands was marketed as an extension of the City it would become just another fringe City area and that those who stepped out of the City, as the 'tiny core' of banking and finance, did so at their peril. Thus, he suggested that Docklands should perhaps accommodate staff of companies with headquarters in the City. However, this policy was based on the expectation, which was later revised, that Docklands would accommodate more firms requiring office and manufacturing space under one roof.

BIBLIOGRAPHY

Barnes, J. (1993) 'State landownership and planning in the spatial restructuring of London's Docklands 1968–90', unpublished Ph.D. thesis, Geography Department, London Guildhall University.

Brownill, S. (1990) *Developing London's Docklands: Another Great Planning Disaster?*, London: Chapman.

Daniels, P. W. and Bobe, J. M. (1993) 'Extending the boundary of the City of London? The development of Canary Wharf', *Environment and Planning A* 25: 539–52.

Docklands Consultative Committee (1990) *The Docklands Experiment: A Critical Review of Eight Years of the London Docklands Development Corporation*, London: Docklands Consultative Committee.

—— (1991) *10 Years of Docklands: How The Cake Was Cut*, London: Docklands Consultative Committee.

—— (1992) *All That Glitters Is Not Gold: A Critical Assessment of Canary Wharf*, London: Docklands Consultative Committee.

—— (1993) *The Urban Regeneration Agency. Property Led or Partnership Led?*, London: Docklands Consultative Committee.

—— (1993) *Community Empowerment in Urban Regeneration: Interim Report*, London: Docklands Consultative Committee.

Docklands Joint Committee (DJC) (1976) *The London Docklands Strategic Plan*, London: Docklands Joint Committee.

Duncan, S. and Goodwin, M. (1988) *The Local State and Uneven Development*, Cambridge: Polity.

Edwards, D. (1985) 'Location perception and investment decision making in London Docklands', unpublished MA thesis, Geography and Planning Department, University College London.

Edwards, M. (1990) 'What is needed from public policy', in P. Healey and R. Nabarro (eds) (1990) *Land and Property Development in a Changing Context*, Aldershot: Gower.

Feagin, J. (1987) 'Local state response to economic decline', in D. Judd and M. Parkinson (eds) *Urban Development in Britain and the US*, Manchester: University of Manchester Press.

Goodwin, M. (1991) 'Replacing a surplus population: the employment and housing policies of the London Docklands Development Corporation', in J. Allen and C. Hamnett (eds) *Housing and Labour Markets: Building the Connections*, London: Unwin Hyman: pp. 254–74.

Greater London Council (GLC) (1982) *The East London File*, London: Greater London Council.

Hansard (1991) *Hansard* 14 February, Vol. 185: 152.
—— (1994) *Hansard* 28 October (Department of the Environment).
Harvey, D. (1989) *The Urban Experience*, Oxford: Basil Blackwell.
Healey, P. and Nabarro, R. (eds) (1990) *Land and Property Development in a Changing Context*, Aldershot: Gower.
Joint Docklands Action Group (1977) *Rebuilding Docklands: Cuts and the Need for Public Investment*, London: Joint Docklands Action Group.
—— (1978) *London's Docks: An Alternative Strategy*, London: Joint Docklands Action Group and Tower Hamlets Action Committee on Jobs.
Ledgerwood, G. (1985) *Urban Innovation: The Transformation of London's Docklands 1968–4*, Aldershot: Gower.
London Docklands Study Team (1973) *Docklands: Redevelopment Proposals for East London*, London: Travers Morgan and Partners.
Marris, P. (1983) *Community Planning and Conceptions of Change*, London: RKP.
Massey, D. (1982) 'Enterprise Zones – a political issue', *International Journal of Urban and Regional Research* 6: 429–34.
Merrifield, A. (1993) 'The Canary Wharf debacle', *Environment and Planning A* 25: 1247–65.
Newman, I. and Mayo, M. (1981) 'Docklands', *International Journal of Urban and Regional Research* 5(4): 529–45.
Travers Morgan (1973) *Docklands Redevelopment Proposals for East London, Main Report*, London: Travers Morgan.

3

LONDON: PLANNING AND DESIGN IN DOCKLANDS

Hugo Hinsley and Patrick Malone

INTRODUCTION

In London, as in a number of cities, redundant docklands have provided opportunities for property capital and new opportunities for alliances between the state and development interests. Until the 1960s, London's East End was a homogeneous centre of working-class housing and employment, dominated economically, politically and socially by the docklands. The port was made up of the powerful forms and hidden spaces of the walled docks. It was remarkable for its scale and for the juxtaposition of domestic life and workers' housing with port facilities and shipping. Although they covered a large area and were located close to one of the world's major financial centres, the docks and the East End were segregated from Central London by inadequate transport links and by a history of social and ethnic discrimination. The East End is still an area of great cultural and ethnic diversity, with relatively high levels of poverty, ill health, unemployment and bad housing. However, it is no longer dominated by the port. Following a succession of dock closures in the 1960s and 1970s, new forms of capital invaded the docklands. Aided by aggressive political policies, the concept of 'Docklands' emerged in the East End as a vehicle for national and global capital.

In many European cities, the railways and ports that facilitated nineteenth-century capitalism have yielded large tracts of development land strategically close to central business districts. London Docklands is archetypal rather than unique. However, due to its size, its location relative to the heart of London and the political context in which it emerged, Docklands became one model of urban regeneration in the 1980s. It reflects changes in the form, mobility and scale of capital and marks a specific era of British post-war urban-planning policy. It epitomizes the politics of Thatcherism, the adoption of a market-led approach to development, the marginalization of planning and the suspension of democratic and legislative controls in favour of a supposed 'free market'. Ultimately, it demonstrates both the rise and fall of monetarism in urban development.

This chapter examines London Docklands in the context of Thatcherism and the changing climate for private investment in urban development. It complements the discussion of the politics and economics of Docklands in the previous chapter and addresses the lessons for planning and urban design.

DOCKLANDS: THE POLITICAL ECONOMY OF INCOHERENCE

The history of Docklands, between the designation of the London Dockland Development Corporation (LDDC) in 1981 and the early 1990s falls roughly into three phases:

(a) the period prior to Canary Wharf
(b) a middle period characterized by the emergence of international interests in Docklands
(c) the period marked by the collapse of Canary Wharf, the dismantling of the London Docklands Development Corporation (LDDC) and the formulation of its 'exit strategies'[1]

Prior to 1986, the LDDC was concerned principally with generating confidence and investment within an open framework for development driven by marketing and tax incentives rather than by planning. Magnetized by public money, Docklands was presented as a zone of opportunity and minimal constraint. It reflected the wider context of 1980s Thatcherism and the Conservatives' efforts to consolidate a shift from the Welfare State and public expenditure to the priorities of the private market fostered by state sponsorship. As Thatcherism increased its influence on Britain, Docklands also symbolized the theory that wealth generated within a 'free economy' would 'trickle down' to engender social and physical improvements in the inner cities (Anderson 1990: 468).

Deregulation took a number of forms as Thatcher attempted to reorder the country to fit her vision of a market-led economy. This was evident in the privitization of public utilities and services, the deregulation of the Stock Exchange in 1986 and the abolition, in the same year, of the Greater London Council and other metropolitan authorities. Against the background of deregulation, the decline of the Welfare State, massive cuts in some areas of public expenditure and the erosion of local government powers and resources, the Conservatives launched new urban development agencies to facilitate private capital investment and the market. The LDDC epitomized these new agencies and their use of public money to fuel private interests in urban development: a model of 'pump-priming' the market with public investment. Docklands, and particularly the Docklands Enterprise Zone, heralded a system of development based on reduced bureaucracy, flexible development controls, prepackaged planning deals and the use of incentives and tax breaks. The Enterprise Zone was a potent symbol of the Conservatives' policy of revitalization by deregulation. Designated in 1982, Docklands' Enterprise Zone covered 192 ha of the Isle of Dogs. For ten years, it offered artificially low land prices, exemption from property tax, income or corporation tax on capital expenditure, from development land tax and industrial training levies. It 'fronted' the redevelopment process in Docklands through the use of a simplified, developer-friendly planning regime and decision-making process.

As a flagship of the government's urban policy, Docklands was to be fuelled partly by reallocating power and resources from democratically accountable local authorities to the LDDC. The affected local authorities lost 245 ha of land and their control over development to the LDDC, but retained responsibilities as planning authorities. They remained responsible for public housing, social amenities, services and infrastructure at a time when financial support from central government for these facilities was steadily reduced. Although the creation of the LDDC required the suppression of local planning powers, it was not constituted as a planning authority with a plan-making brief. The LDDC was given a dual role. First, to provide land and infrastructure for development. Second, to market Docklands and to alter perceptions of the area through imagery based on 'wind surfers, power boats, swans, spectacular sunsets and a few isolated pockets of "heritage" that still survived' (Davies 1987: 31).

The government and the LDDC viewed Docklands as a *tabula rasa* and rejected strategies for regenerating the area that were proposed under the 1973–79 Labour government (see Chapter two). The LDDC also rejected the ethic, embodied in earlier plans, that Docklands should attempt to provide directly for the needs of the 40,000 people who lived in the area. The policy was that private capital was to be attracted into commercial development in Docklands on its own terms, with its passage being eased by public money channelled through the LDDC and the Enterprise Zone. The Thatcher government was clear in its priorities for Docklands. Planning, in any social or strategic sense, was seen as a deterrent to private investment. Local politics and a concern for local needs were held to be against the national interest. The primary aim of the LDDC was to establish market confidence. Planning became a sales-based, marketing tool. In 1982, Reg Ward, the LDDC's first Chief Executive, stated:

> We had to hype-up the place [Docklands] or the financial institutions would not have come . . . in the public sector so much time is wasted in creating uncertainty. This is not what happens in the real world. We have no land-use plan or grand design; our plans are essentially marketing images.
>
> (quoted in Hatton 1990: 61)

Ward's equation of the 'real world' with property and financial institutions,

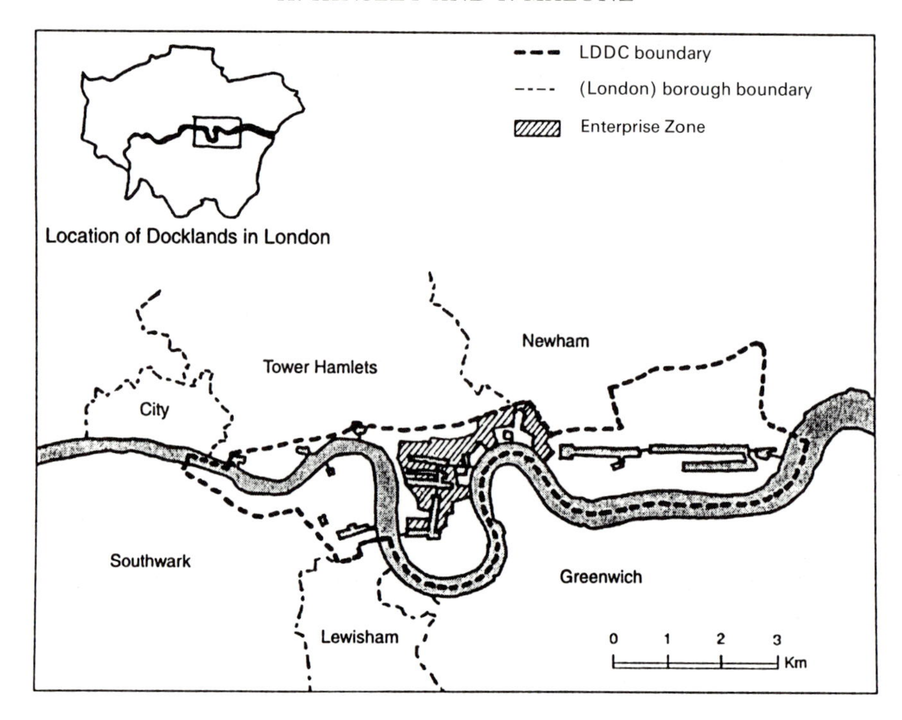

Figure 3.1 London Docklands: location map

and his scornful attitude to planning, were typical of government policy in the 1980s. Against the background of social unrest and riots in British inner cities, the LDDC broke with post-war planning. It was explicitly opposed to a strategic planning framework for Docklands and followed the government's directive that urban-development corporations should 'do things' and be 'free from the inevitable delays of the democratic process' (Michael Heseltine, quoted in Thornley 1991: 181). The LDDC did, however, split their area into four zones and had a crude land-use proposal for each zone (Fig. 3.1). For example, the Isle of Dogs was designated mainly for commercial development stimulated by the policies of the Enterprise Zone, while the Surrey Docks were to be a major area for new private housing.

The use of a *laissez-faire* approach to urban development reflected the general tendency towards deregulation in the economy. It followed from the government's intention to give free rein to British development interests, while drawing capital into Docklands from within a competitive, global economy. In this respect, Docklands mirrors the relationship between deregulation in planning and the efforts of governments and urban administrations to attract mobile flows of capital into their territories – a process of seduction pressurized by urban marketing and the competition between 'world cities' and financial centres. Thus, while the LDDC's 'fly-paper' approach to development lured investment through the use of financial incentives and the suppression of planning restrictions, deregulation dictated that the primary purpose of the LDDC was to establish confidence and interest in Docklands. Public money used for marketing, infrastructure and incentives was intended to make Docklands attractive to developers. This policy of seduction was justified by the LDDC on the basis that developers were not interested in Docklands. It was argued that 'at best the Isle of Dogs could become a light industrial park [while there was little] potential for development in the Royal Docks Area' (Attwood 1989: 121–2). On the basis that there was no measurable force for redevelopment, it was held that the 'ultimate destination' of Docklands could not be predicted and, therefore, that traditional planning mechanisms could not be applied. In short, it was argued that planning was impossible and (in the first years of the LDDC's reign) that urban design could exist only as 'a marketing tool aimed at unlocking the potential for regeneration of the area' (Attwood 1989: 121).

While it was used to support the rejection of planning and urban design, the LDDC's pessimistic view of Docklands' prospects contrasted sharply with the optimistic marketing imagery that it created to generate development activity and investment (Fig. 3.2). This contradiction provides an

Figure 3.2 London Docklands: Jean Michel Jarre's 'Destination Docklands' concert, 1988
Source: London Docklands Development Corporation

interesting insight into the ethics of the LDDC in the early 1980s. One explanation for its apparently illogical position on Docklands' prospects is that it perceived development opportunities to which the development industry was blind. Another is that its pessimistic assumptions about development interest in Docklands were used to justify an innate urge to marginalize planning.

Whatever the 'logic' of the LDDC's position in the first half of the 1980s, its position on planning and urban design altered with the invasion of Docklands by large international interests in the mid-1980s. The arrival of the Americans and the emergence of the Canary Wharf project marked a turning point, particularly for the Isle of Dogs. The growth of foreign interest might be seen as vindicating the LDDC's initial policies. But it might also be seen as part of a general trend in British property markets in the middle and late 1980s. Foreign investment in property in the United Kingdom in 1988–89 reached £3.1 billion, with Japanese companies accounting for 40 per cent of that total (Budd and Whimster 1992: 271).

In 1985, the American developer G. Ware Travelstead began to talk up the investment potential of a large-scale development at Canary Wharf. It was argued that the needs of the deregulated financial markets could not be met within the City of London except through the extensive destruction of its historic fabric, and that the 'big bang' might fuel an eastward extension of London's financial core. Travelstead proposed a development of 10 million ft^2, comprised mainly of office space and dominated by three sixty-storey towers. However, to the embarrassment of Mrs Thatcher, whose political persona had become inextricably linked with Docklands, Travelstead could not raise financial backing and withdrew from the project in 1987. Possibly spurred by the imminent general election, the LDDC approached Olympia and York as a company of sufficient size and financial muscle to undertake a development on the scale of Canary Wharf. In July 1987, buoyant with its success in the early phases of New York's Battery Park development, the company took on the project, together with an attractive package of subsidized land, tax incentives and political patronage. Thatcher and the LDDC were in a hurry to produce visible results. The company had no time to rethink Travelstead's project, and took over the original master plan. This plan had been approved by the LDDC in only fifteen days. In giving this approval the LDDC had not consulted the GLC or the relevant Docklands local authorities. Nigel Spearing, a local MP (Newham South), expressed the general surprise and disgust at this:

> Possibly the largest commercial project in the world was subject to less review and scrutiny than a planning application for an illuminated sign on a fish and chip shop in the East India Dock Road.
>
> (Docklands Consultative Committee 1992: 9)

The proposals for Canary Wharf in 1985 symbolized the ascent of Docklands as a magnet for global capital. Political and economic perceptions changed as the Isle of Dogs emerged to compete with the City of London and with other cities as a potential office centre in the global economy. The spirit of rather low-key opportunism, exercised within an open and relatively chaotic framework for development, gave way to an era of more sophisticated exploitation and international interests. Land prices in the Isle of Dogs increased from £80,000 to £250,000 per acre in the five years to 1986. The earlier model of random, incremental, mixed development was pushed aside as Canary Wharf materialized as the commercial heart of the LDDC's 'water city of the twenty-first century'.

It could be argued that the LDDC failed to anticipate the strength of foreign investment in Docklands and that the emergence of Canary Wharf shed doubts on the wisdom of its initial policies. Its 'hands-off' approach to planning and urban design was shown to be inadequate and misconstrued in the light of events in the second half of the 1980s. First, because the large international interests which entered Docklands in the mid-1980s demanded planning and urban design strategies from the LDDC in order to reduce development risks and enhance development profits, and second, because in the early 1980s the LDDC and developers failed to see the Isle of Dogs' potential to accommodate higher economic uses at greater densities. Although designated by the LDDC as an office and commercial core, the first developments on the Isle of Dogs followed the pattern of small-scale 'decorated sheds', which were adaptable for office and light industrial uses. Thus, by the mid-1980s, some new buildings and proposed developments were already redundant in terms of land values and the potential for development profit.

By its scale, relative coherence and the quality of its constructional and development processes, Canary Wharf brought a new dimension to Docklands. As envisaged by Olympia and York, it was to provide 40,000 jobs and had a projected development cost of £4 billion. Comprising twenty-six buildings, mainly designed by American architects, it was to accommodate 10 million ft^2 of office space, 200 shops, restaurants, cafés, health clubs – but no houses. Rather than present the project in a competitive light in terms of the City, Canary Wharf was said to offer a 'logical extension' to London. It was intended to complement the City as a financial centre, while physically reflecting the architecture of one of the 'world's great cities' (Budd and Whimster 1992: 237). Whether it can be seen as an extension of London's commercial core is open to question. It did, however, distend London's collapsing property markets.

The success of Canary Wharf was undermined by a downturn in the demand for office space, although Olympia and York clearly expected to survive a property recession. Canary Wharf was also undermined by the financial and bureaucratic powers of the City of London which reversed a conservationist approach to development resulting in a 30 per cent increase in the total stock of office floor space in the City between 1979 and 1989 (Budd and Whimster 1992: 23). It has also been estimated that 30 million ft^2 of new office space was built in the City of London between 1985 and 1993. This new space represented 50 per cent of the stock of office space in the City, although, because much of this replaced existing space, the net addition in total office floor space amounted to roughly 12 per cent (Cassidy 1992: 9).

Initially, it was predicted that the demand for office space would be such that Canary Wharf would inflict no net losses on the City's employment and office markets (Budd and Whimster 1992: 235). This assumption was rejected by the City's financial and property lobbies, which, faced by the threat of Docklands, reversed policies that favoured conservation and the suppression of development activity. Within the City, developers were licensed to increase office densities and moves were made to identify new sites for office development. The prospects for Canary Wharf were challenged by the City's massive Broadgate office development and by a rash of new 'groundscrapers' centred mainly on railway stations in and around the City. Within eighteen months of reversing its policies on development, the City approved 20 million ft^2 of office space. Thus, competitive conditions were emerging in London's office market even before Canary Wharf started on site in 1988. Moreover, there were signals that growth and demand in the financial sector might be impeded.

The difficulties facing Canary Wharf were exacerbated by the failure to link the project successfully into London's transport network. In this respect, Canary Wharf suffered the effects of a general neglect of transport issues in Docklands. Transport should have been a key element in the integration of Docklands with the City and the rest of London. The LDDC's programme implied the movement of large numbers of people. Canary Wharf's proposed 40,000 workers were equivalent to roughly 13 per cent of the City of London's work-force (Budd and Whimster 1992: 23). However, the LDDC's non-planning approach to transport meant that only token gestures were made towards strategic transport planning – a neglect underpinned by the government's antipathy to public transport. The government did not agree to begin construction of the Jubilee Line to extend London's Underground into Docklands until twelve years after the creation of the LDDC. In the meantime, the absence of a lifeline into the Underground system impeded the regeneration of Docklands. Initially, the government adopted the position that private investors should contribute financially to Docklands transport infrastructure. Later the government was trapped between an inability to back down and the realization that good transport, and particularly the Jubilee Line extension, were necessary to Docklands. For their part, Olympia and York had agreed to contribute £400 million to the Jubilee Line over twenty-five years. But as it slid towards bankruptcy in 1991–92, it was unable to pay the first instalment of £40 million. Consequently, the development of the Jubilee Line was further delayed.

Daniels and Bobe (1993) argue that the development of Canary Wharf ignored the traditional concerns of the development industry regarding location. They claim that, buoyed by assumptions regarding demand, financial incentives and the late-1980s ethos of the property and financial sectors, Canary Wharf set aside considerations of: location, infrastructure, accessibility, labour market, support services or prestige (Daniels and Bobe 1993: 541). At 3.5 km from the nearest point in the City, they argue that Canary Wharf is not as well located as The World Financial Centre (New York) or La Défense (Paris). Thus, the project was highly dependent on the creation of an island of office space with a sufficient critical mass, on the quality and services offered by its buildings, heavy marketing and the visual impact of its 800 ft tower.

Given the difficulties faced by Canary Wharf, it could be argued that Olympia and York allowed itself to be caught up in the hysteria of the mid-1980s and the 'big bang'. Tempted by huge tax subsidies and a low land price, the usual developers' priorities of location and access were set aside. By 1991, however, the company was calling for a fifteen-year planning and transport strategy and was openly critical of the government and the LDDC for producing 'an urban failure' (Barrick 1991: 1). But Olympia and York was not alone in misjudging the prospects for Docklands. In the late 1980s, deregulation and a property boom generated a number of large projects and an attempt at 'Manhattanization' which gave rise to visual chaos and a spate of market failures on the Isle of Dogs. In 1992, 53 per cent of the new office space generated by this process was vacant. Between 1988 and the early 1990s, office rents fell from £60–70 to £30 per square foot. As rents dropped and significant inducements were offered to potential occupiers, much of Docklands' office space fell into the hands of receivers. Canary Wharf, standing bankrupt and half empty on its island of corporate space, became a monument to the failure of property markets.

SOCIAL CONSEQUENCES IN DOCKLANDS

The LDDC's programme proposes an increase of 75,000 in Docklands' population, that is, from 40,000 to 115,000 in the period 1981–2014. It anticipates a rise in the number of private dwellings from roughly 5,000 to 36,000 units. It also calls for an increase from 27,000 to 200,000 jobs – many to be accommodated in 65 million ft^2 (6.0 million m^2) of new office space. By 1994 the population of Docklands was 65,000, well below the target of 115,000 set for 2014. Roughly 13,500 of the proposed 36,000 private housing units had been completed, and about 20 million ft^2 of the proposed new office space. Thus, given the ambitions of the LDDC, the rate of development in Docklands is lower than might be expected and development is currently retarded by a prolonged recession in the property markets. However, there has been a remarkable change in Docklands and, arguably, it is not the volume of development that is at issue but rather the social, economic and physical results of the LDDC's programme (Fig. 3.3).

Figure 3.3 London Docklands: 'Doglands'
Source: Louis Hellman

Socially, Docklands epitomizes the deepening divisions in British society. It echoes Disraeli's 'two nations':

> between whom there is no intercourse and no sympathy; who are ignorant of each other's habits, thoughts and feelings, as if they were dwellers in different zones or inhabitants of different planets.
>
> (Disraeli 1881)

In the 1980s, Docklands mirrored the division between an emergent 'yuppie' culture centred on the financial world of the City and the neglected, economically redundant, urban poor. Loaded in favour of private capital, the LDDC's programme has done little to benefit the indigenous population in terms, for example, of employment and housing. Local unemployment has increased since 1981 and roughly 75 per cent of the 'new' jobs in Docklands have been transferred from other areas. Some menial jobs are available to the local population, but training programmes have had little effect in preparing local people for office employment (DCC 1991). Similarly, the LDDC's output in terms of private and owner-occupied housing contrasts with the levels of homelessness and deprivation suffered by the local population (see previous chapter).

Docklands' divisive social geography is reflected in the LDDC's housing policies, which have transformed social conditions and triggered conflicts between the indigenous and new populations. Between 1981 and 1989, the proportion of dwellings in owner-occupation increased from 5 to 44 per cent. This was a radical change, particularly when contrasted with the deteriorating conditions for public housing. Docklands' residents must contend with the frustration of living in a huge construction site and a lack of good public transport, shopping facilities and other services. In addition, the residents of public housing suffer the effects of the LDDC's neglect of and the government's withdrawal from the public-housing sector.

The LDDC was in place for eight years before it confronted the issues of public rental housing and 'community services'. Its initial views on social issues were expressed by its first chairman (Nigel Broakes) who held that the LDDC was 'not a welfare association but a property-based organization offering good value' (quoted in Ambrose 1986: 228).

As the product of market-led development, the restructuring of Docklands provides privileges which may be secured with rent or purchased through the housing market. Those privileges are denied to those who cannot gain access to Docklands' housing and may be granted at the expense of the indigenous population, for example where local people are denied access to a privatized waterfront. Isolated within what was a predominantly working-class area, the residents of the tower blocks on large public housing estates have looked on the transformation of the Docklands with rising frustration and anger. The resources to repair and upgrade public housing have been curtailed, while massive public funding has been invested in a failing vision of the future that has little to offer the original resident population. Underlying the instances of local anger, arson and vandalism, there has been a steady build-up of dissatisfaction within local communities. The promises of new opportunities, jobs, training, better health care and education have not been realized after fourteen years. Feelings of alienation and divisiveness have spilled over into a series of battles between the LDDC and the local population. In 1992, after years of sweet-talk from the LDDC and major development companies, the residents of Poplar and the Isle of Dogs acted to sue the LDDC and Olympia and York for the disruption caused by development. By acting together, residents aimed to gain damages of £100 million.[2] But dissatisfaction and housing shortages also fuelled racism and local support for the British National Party (BNP). Housing issues were a key element in the election of a BNP local government councillor for Docklands' Millwall in 1993.

At another level, the local authorities have waged war with the LDDC. Newham Council, for example, has been in conflict with the LDDC because 'planning-gain' agreements have failed to deliver housing and social benefits in the Royal Docks. This conflict is typical in that the LDDC, having raised expectations, would deliver community benefits only 'through the market' and specifically through land sales. The conflict is also typical in that the LDDC is now trying to market the Royal Docks to developers using the concept of a private sector 'urban village', while the local authority has drawn up a comprehensive master plan for the Royal Docks which is underpinned by social housing.

ECONOMIC GOALS

At the heart of Docklands, Canary Wharf and the Isle of Dogs provide a focus for the conflict between Britain's 'two nations'. While it is clear that the Docklands redevelopment has failed to meet the needs of the local population, the government and the LDDC have also failed the property markets and the political and economic interests that government policies were intended to serve. Whereas Canary Wharf is an obvious symbol of the failure of the commercial property markets, the collapse of the housing market in the late 1980s was felt keenly in Docklands. In 1990, about 4,000 of the 13,500 new private housing units built in Docklands were unsold. As the market collapsed and house prices fell, purchasers who entered Docklands during the housing boom of the 1980s were saddled with onerous mortgages. Some were faced with repossession or 'negative equity'.[3]

In terms of commercial development, Docklands has shown that market-led development can be fed by factors of supply rather than demand. The combination of state-sponsored tax and other incentives, relatively high levels of investment pressure and low levels of demand can overheat markets and induce artificial, investment-led booms. Docklands demonstrates how supply-led processes may feed development cycles in a frenzy of speculative development, which results in falling rents and capital values, bankruptcies and political embarrassment. The dramatic failure of Canary Wharf and other projects are very visible signs of the dangers of over-extension in a supply-led market.

The relationship between Docklands and the market can also be examined in terms of the failure of the leverage or 'pump-priming' process to trigger sustained private investment. Docklands has absorbed around £5 billion of public money since the LDDC took over in 1981, but it has attracted only £9 billion of private investment.[4] Despite the rhetoric of monetarism, Docklands is state led in the sense that it has been heavily sponsored by the government. Aided by a lack of public accountability and weak democratic processes, the alliance of political and private interests in Docklands has resulted in ever-increasing public expenditure to support flawed planning and economically flawed development processes. The deal between the government and Olympia and York exposes the fallacies of 'leverage', in that the contribution which Olympia and York was to make to the Jubilee Line (£400 million), together with its other commitments, were paltry if assessed in terms of the subsidized land price paid by the company for the site for Canary Wharf and Enterprise Zone tax concessions estimated at around £1.3 billion. The company's contribution to the transportation network was little more than a political fig-leaf in comparison with the state's sponsorship of Canary Wharf.[5]

PLANNING AND URBAN DESIGN IN DOCKLANDS

The marginalization of planning and urban design, the LDDC's policy of packaging dislocated areas of development land, and the alienation of local authorities, have combined to create a process of uneven development in Docklands. With the relaxation of planning controls, development activity has taken root on the basis of relative accessibility and profitability. The bulk of investment has flowed into the areas nearest to the City of London, although these areas had less need of the privileges bestowed by the LDDC and the Enterprise Zone. Beyond the Isle of Dogs and Canary Wharf, the huge territory of Docklands contains a range of different areas that generate different perceptions and disappointments. The Isle of Dogs may appear to be an urbane fantasy when viewed by tourists from the robotic train but, to the east, Docklands offers an extension of suburban Essex rather than of urban London.

Docklands is developing as a patchwork of dislocated and predominantly monofunctional areas. Its enclaves of office and commercial space and various housing developments do not connect within a coherent urban structure. The area is spatially, architecturally and functionally disjointed. There is some good architecture and some credible urban design, but a relatively small number of creative projects are lost within a desert of chaotic urban space. Docklands is proof that a development process driven by land packaging and marketing promotes incoherence and disorientation, whether in the 'office city' on the Isle of Dogs or in the speculative suburban housing that carpets much of Surrey Docks and Beckton. In Surrey Docks, for example, low-density housing has been spread over and around the reclaimed docks. A ring road feeds the culs-de-sac and parking courts of this 'newtown-in-town'. Families without a car are stranded and, because of low population densities, shopping facilities are minimal. Developers bidding for sites without the guidance and support of a comprehensive plan have opted for a safe approach and reproduced well-tried housing models. Thus, suburban development occupies land which is within two miles of the City. The physical and social structures of suburbia contrast with 'luxury' apartment blocks protected by gates, TV cameras and guards. Some of Docklands' housing makes a real contribution to architecture and urban design, but individual efforts to achieve an appropriate scale and design quality sit alongside vast areas of banal development.

The incoherence of Docklands' physical and land-use structures has been underpinned by the LDDC's failure to develop an adequate transport infrastructure in line with the pace and scope of redevelopment. This failure has generally impeded development, and false signals about transport infrastructure have contributed to bad and sometimes unsustainable investment decisions. The elegant Tobacco Dock, converted into a shopping arcade on the basis that the Docklands Light Railway would service the development, is now bankrupt and largely vacant. Difficulties have befallen the London Arena, a large venue for concerts and events, but developed without convenient public access. Further from the 'honey-pot' of the City, the Surrey Docks and Royal Docks also suffer from a lack of transport links. Access is a major problem in the eastern reaches of Docklands, an area that has at its heart the largest enclosed docks in the world, with 90 ha of water and 220 ha of wharfage. In the heady atmosphere of the 1980s property boom, two large private consortia floated projects for this area, which included a huge entertainment complex and the biggest shopping centre in Britain. These projects faded away, partly due to doubts about the lack of a public transport strategy.

The problems of Docklands, in terms of transport, planning and urban design, reflect the paucity of the LDDC's strategies in the first half of the 1980s. In this period, planning was particularly excluded from the LDDC's agenda. It would be wrong, however, to suggest that British urban designers were entirely barred from Docklands prior to the mid-1980s. As early as 1981, David Gosling, Gordon Cullen and Edward Hollamby (chief architect–planner of the LDDC) prepared urban design proposals for the Isle of Dogs. The ensuing report (*Isle of Dogs: A Guide to Design and Development* (1982)) was dominated by Cullen's 'townscape' approach to urban design. Cullen's proposals allowed for a 'necklace' of communities on the periphery of the Isle of Dogs, with major commercial developments surrounding the interior water basins. However, Cullen eschewed the idea of an overall plan and, in retrospect, his imagery and perspectives have been described as 'utterly naïve' and 'preposterously aesthetic' (Davies 1987: 31; Buchanan 1989: 42). Cullen's proposals hinged on the optimistic assumption that the LDDC would make a significant investment in Docklands' public realm, in its open spaces, streets, squares, and public-transport systems. He held that development controls and investment in the public realm could provide a framework for the buildings, which would define and enclose public space.

In contrast, Gosling operated at a spatial level above that of Cullen. Gosling's initial studies in Docklands were concerned with major vistas and axes. He dismissed Cullen's approach as being unlikely to provide a coherent spatial structure, and has consistently argued for an urban-design framework based on an axial spatial structure – although his approach to urban design is essentially 'megastructural' rather than *beaux-arts*. In 1989, he summarized this approach to Docklands as follows:

> An urban design plan may be regarded as a plan for the public realm. The public realm is the skeletal structure of the city . . . [it] includes the movement systems of the city; the highways, streets, pedestrian routes and networks, as well as rapid transit systems. But it also includes the public spaces, squares, parks, piazzas and arcades.
>
> (Gosling 1989: 116)

In 1985, the gap between the LDDC and the urban-design 'lobby' was expressed by work on the Royal Docks, undertaken by students working with Gosling and Michael Wilford. This work also stressed spatial factors as a framework for development. It played on the use of the urban block, axes, vistas, landmarks and historical 'tendencies' in Docklands spatial structure (Fig. 3.4). As Gosling's complimentary references to Leon Krier suggest, this work reflected general tendencies in urban design in the 1980s. It displayed the common weaknesses of a formalist and spatially based approach to urban design. *Beaux-arts* axial planning, it was argued, offered spatial and historical continuity and a strong structure on which different land uses and individual developments could be supported. But this approach did not acknowledge the close relationship between urban form and land use.

Urban designers have been concerned predominantly with formal and spatial issues in Docklands. Whereas Gosling and Cullen differed on the need for a master plan, both emphasized the public realm as a basis for development. Their proposals were based primarily on the spatial or physical structure of Docklands rather than its land-use or functional structure, and they laid little stress on infrastructure or the pressures exerted by development interests. In emphasizing formal issues, they may have missed

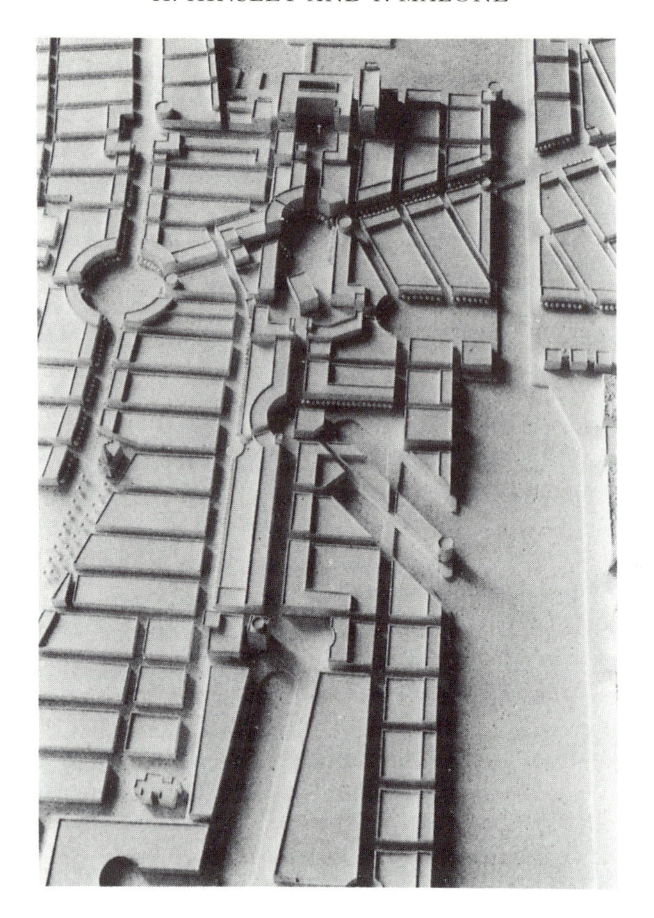

Figure 3.4 London Docklands: massing study: Royal Docks 1985
Source: Stephen W. Proctor

opportunities to provide strategies for Docklands based, in greater measure, on planning, land-use and market criteria – that is, strategies that might have been integrated with those of the LDDC. Thus, it can be argued that the form-based strategies of the urban designers were challengeable in themselves and unlikely to penetrate the defences of market-led development interests. This also implies that the absence of good urban design and planning in Docklands followed in part from the intransigence of the LDDC and partly from weaknesses in urban design theory and practice.

The LDDC rejected the conclusions of Gosling and Cullen as being too restrictive. Charles Attwood (Principal Architect of the LDDC) explains the conflict between the urban designers and the LDDC in terms of the reluctance 'of some urban designers' to recognize the need for an introductory phase in the development process; that is, a phase in which planning is marginalized in favour of development interests and the promotion of confidence. Against those who believed that 'a more fully developed urban plan was possible at the very start' Attwood argues that:

> given the market and development conditions in which we operate in the UK, the initial incremental approach to building up development momentum is unavoidable, and [that] any urban designer who fails to understand the need will be of limited value.
>
> (Attwood 1989: 122)

Attwood holds that the 'ultimate destination' in Docklands could not be established prior to the emergence of a market for Docklands. He highlights Canary Wharf as a watershed and suggests that planning and urban design had to be marginalized, or subjugated to marketing, until such time as the property industry decided the direction that should be taken in the redevelopment process.

Contrary to this view, it might be argued that the emergence of international interests in Docklands exposed the weaknesses of the LDDC's vision and the costs of its rejection of planning. Transport systems were shown to be inadequate and land in the Isle of Dogs proved to be underdeveloped and undervalued. Moreover, given the ambitions of international interests, it became clear to the LDDC, much too late, that planning and design were essential to attract major developers and global capital. The LDDC adopted the view that planning and urban design are factors which can enhance development profits and reduce development risks. With the influx of international interests into Docklands, an essentially weak approach to planning and design emerged to fill the vacuum in the LDDC's agenda. This approach was opportunistic, form based and limited in terms of its capacity to provide a comprehensive framework for development. The consequences became evident as Canary Wharf emerged as a 71-acre 'island' of international capitalism.

While Canary Wharf is only a part of Docklands, it is symptomatic in that it was initiated without a comprehensive planning framework and is based on a limited urban-design strategy. This is reflected by the manner in which the project is plugged into London's spatial, functional and transport systems. It is moored off the edge of the commercial core, although its form and style are intended to convince potential occupiers that it is a settled part of London (Figs 3.5 and 3.6).

Canary Wharf is expensively dressed in granite, marble, York stone, brass, London plane trees and British oaks (airlifted from Hamburg). The product of computer-generated imagery, the project has been described as a 'ready-made and pre-painted kit of avenues and circuses, street lamps and railings, benches and paving stones [dressed in] Oxford shoes and a London

Figure 3.5 London Docklands: a view of Canary Wharf from Greenwich
Source: Olympia and York

Figure 3.6 London Docklands: a view of Canary Wharf from the west
Source: Olympia and York

club tie' (Glancey 1991: 27). Its modern, fifty-storey, 800 ft stainless steel tower suggests new technology, while its neo-London base draws on images of the past. Its historicism clothes a steel structure built by fast-track construction techniques. Its *beaux-arts* ground plan was created by a Chicago-based practice (Skidmore Owings and Merrill), which is better known for its 'modernist' commercial architecture. As such, Canary Wharf might stand as a paradigm of late twentieth-century architecture, which exploits historical imagery created in the arena of international capital. The combination of a sleek, modern office tower, neo-Edwardian base and *beaux-arts* landscape may be perceived as bizarre, but it expresses the reality of commodification and the alliance of aesthetics and rent. Socially, Canary Wharf is a place of corporate imagery. The historical references are to public urban space, but this is private territory produced for private ends.

In the late 1980s, the approach to urban design in Canary Wharf was spread into adjoining areas. Plans were made for further high-density development, based on the application of aesthetic controls to buildings rising from a fixed ground plan. Much of the architecture that followed from this approach offered little optimism for the future of design in Docklands. But

Figure 3.7 London Docklands: West Silvertown, an urban village in Royal Docks
Source: London Docklands Development Corporation

urban design did gain a foothold in Docklands in the late 1980s. Some credible urban design has emerged, although this is confined to small areas and isolated projects. An 'urban design team' was incorporated into the LDDC, but this was disbanded in 1991. The architect, Richard Rogers, prepared a development framework for The Royals in 1992, and further proposals followed for the Isle of Dogs, Limehouse and Leamouth in 1994. The 'urban villages' movement has also surfaced in Docklands (Fig. 3.7). In 1989, aware of much criticism, the LDDC set up an advisory 'design panel' of architects and urban designers. However, this can do little to redress the sense that planning and urban design have remained marginal to the LDDC's interests, and the recent initiatives of consultants and others have produced few real benefits on the ground.

Efforts to raise the profile of planning and design in Docklands are dogged by low levels of development pressure and by the LDDC's failure to install an adequate planning policy at the outset. But Docklands also begs the question of whether planning and urban design professionals are up to the job of formulating successful policies for large-scale projects. This is an important question to raise in the context of Docklands and the current era of urban development. It could be argued that planning has been marginalized while architects and urban designers have been employed to generate frameworks for development. Thus, while planners have been largely excluded, architects and urban designers have had a bigger role in the development of Docklands. This prompts the conclusion that Docklands has exposed the inadequacies of the form-based policies of architects and urban designers as a basis for large-scale development, and that approaches to urban development derived within architecture need either to mature and/or be more closely linked with planning.

CONCLUSIONS

Docklands highlights two paradoxes of British urban policy in the 1980s. First, that the rhetoric of deregulation and the free market was backed by massive state intervention and public subsidies. The state manipulated decision-making and planning processes, land values and property markets, while promoting private capital through systems of state sponsorship. Second, Docklands shows that the positive imagery of profitability and market freedom, as proclaimed by the government and the LDDC, was spurious. The new economic order has brought social fragmentation and division. It has done little to benefit the indigenous population of Docklands and has also damaged some of the property and allied financial interests which it was intended to benefit. Moreover, in Docklands, it has provided evidence of the physical costs of deregulation and the marginalization of planning.

The LDDC now admits that there are problems in Docklands, although its public-relations office has created the phrase 'the problems of success'. Apologists for Docklands argue that its problems are mostly due to the speed and scale of change and that, given time, the area may make a great contribution to London (Hall 1992: 11). As the LDDC approaches the end of its term, its Chief Executive is not repentant:

> We want the LDDC to be a considered, organised landowner and development control authority, but we are not going to impose on developers. I have no regrets about Canary Wharf at all. I think Docklands was a fairly well controlled area. Problems in London are not a planning issue – they were caused by availability of finance, excessive optimism and deregulation of financial markets.
>
> (Eric Sorensen, quoted in *Building Design*, 6 August 1993)

There have, however, been changes in the government's urban policy in the 1990s. For example, after stinging criticism from a House of Commons Select Committee on Employment (HMSO 1988) the government has now accepted the need for education and training, although remedial action has been frustrated by a prolonged recession and by the persistence of Thatcherism. Similarly, the LDDC now speaks of community development, but this late and partial conversion to community interests is not backed by significant resources, and clings to the ideology of the market.[6] Increasingly, the LDDC falls back on the excuse that urban development corporations were not intended to solve broad social or planning problems. Beyond the LDDC, however, there is a growing recognition that market-led development carries economic and social costs. Public investment is devalued. The social problems of unemployment and homelessness damage the individual, society and the economy. The rejection of planning and urban design undermines profitability and generates development risks.

Capital has sought to exploit Docklands without allegiance to social or planning frameworks. At best, planning has been defined as a marketing tool, or has been used to exploit the relationship between aesthetics and rent. But Docklands has also hosted a major disaster in property speculation and banking. It has demonstrated that the market-led approach to development is flawed if assessed on its own terms, as well as in terms of the extraordinary opportunity afforded by Docklands to remake a large part of London.

The failure to address broader social, economic and design issues is a massive indictment of a development policy that characterized the 1980s. Docklands bears the scars of a market-led approach to urban regeneration. Its physical and functional structures, its transport system and the failure of its property markets attest to the costs of market-led development. The Isle of Dogs is remarkable in terms of the quantity of space developed, but not in social, economic, urban and architectural terms. Similarly, the quality of much of the redevelopment throughout the LDDC area is mediocre to the point that Docklands can be interpreted as a major lost opportunity for urban planning and design in London.

It is difficult to identify the logic of abandoning planning controls as an incentive for development, when 'open planning' fails to inspire confidence in developers and leaves the development authority without an overall plan. The LDDC's position hangs on Attwood's argument that an initial period of 'market building' was essential to Docklands, and that this required the exclusion of planning and the subjugation of urban design to marketing processes. However, given that the initial period in Docklands' development set a pattern of incoherence for the entire project and provided an unstable basis for investment, it is difficult to see the logic of the LDDC's policies; except insomuch as these policies expressed a wider political agenda and the government's urge to reduce Britain's democratic planning machinery.

However, while the self-defeating nature of the LDDC's policies provides one of the lessons of Docklands, other conclusions may be drawn which have implications for urban design and planning. A great deal of criticism has been levelled at the political, administrative and economic processes that engendered Docklands. Much of that criticism has been justly directed at the government and the LDDC. However, some responsibility for the failings of Docklands may be laid at the door of planning and, particularly, the design professions. In detailing the alternatives open to the LDDC, frequent references are made to the social values (however compromised) of British post-war urban policy. Little reference is made, however, to the capacity of professionals to produce well-designed environments, or to their ability to provide satisfactory alternatives to the LDDC's policies. It would be unwise to assume, for example, that the planning regime that existed in the docklands prior to the LDDC could have delivered good physical planning and urban design. Neither is it clear that, given a more sympathetic hearing, the British planning and design lobbies could have delivered appropriate and successful strategies for Docklands.

Although the view of planning and design adopted by the LDDC was obviously unsympathetic (or exploitative), it could be argued that planners and urban designers failed to overcome the LDDC's resistance to planning. Moreover, Docklands may have exposed essential weaknesses in the theory and practice of form-based approaches to urban design. It demonstrated that the responses of the professionals to Docklands were weakened by divisions within and between the planning and architecture lobbies. As Davies points out, many architects found satisfaction in the snub delivered by the LDDC to the planning profession, on the basis that, in the eyes of many architects, 'town-planning legislation has done nothing positive for the environment in British towns and cities' (Davies 1987: 32). Thus, whereas Docklands may be seen as a lost opportunity for planning and urban design, the entire responsibility for Docklands should not be laid at the doors of the government and the LDDC.

As the political and economic frameworks of the last decade fade into history, valuable lessons may be drawn from Docklands in terms of the future for planning and urban design. Thatcherism is now a diminishing factor in British politics. The LDDC is being wound up. The present government has lost interest in Docklands, and private capital is struggling in a recession. Even a government so uninterested in strategic planning cannot ignore the implications of the Channel Tunnel (finally operational in 1994) and has turned its attention towards the concept of the East Thames Corridor. This concept was first proposed by the South East Regional Planning Conference (SERPLAN) in 1990 and was adopted by the government in 1991. Later renamed the Thames Gateway, the 'corridor' starts from the Royal Docks and Greenwich and links roughly 4,000 ha of development land, stretching 40 miles eastward to the sea (Fig. 3.8). The area has a population of around 1.6 million and is overseen by sixteen elected local authorities as well as the LDDC. The use of one large development agency akin to the LDDC has been ruled out in favour of collaboration between local authorities and a government 'task force'. The government's thinking about Docklands regeneration was driven by ideology and by an assumption of ever-expanding growth: very much a 1980s belief. The 1990s are different – a time of stasis and retrenchment, and with a much less confident government.

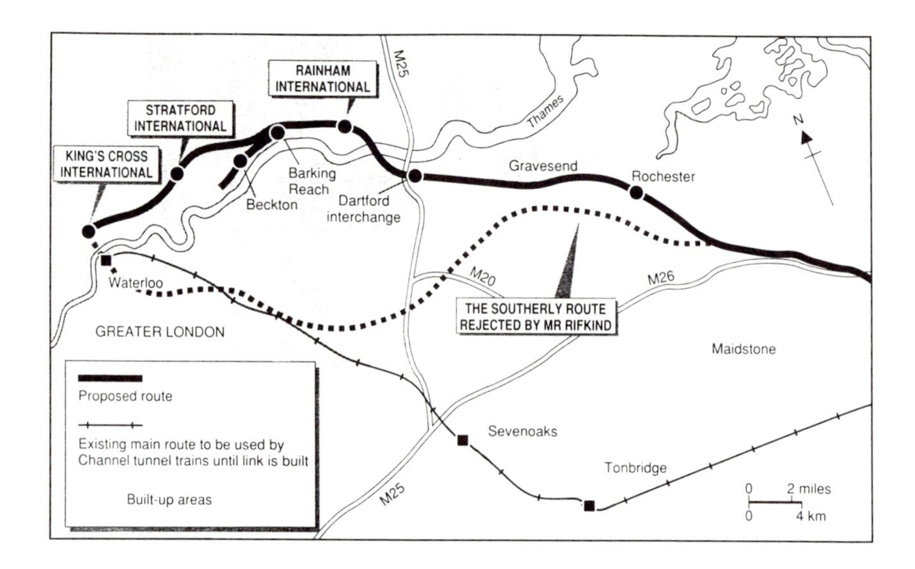

Figure 3.8 London: Thames Gateway
Source: J. Hall 1992

It seems likely that the Thames Gateway will repeat the mistakes of Docklands in terms of the failure to evolve planning strategies, and to invest sufficient public money in transport, infrastructure and a framework for development. The high-speed rail link from the tunnel to London is clearly one key to the regeneration of the area, but the government has stalled for years about its route and its financing, and it is uncertain when it will be completed. While the government focuses its hopes for London's development on the Thames Gateway, it has disabled the mechanisms for broader strategic planning in London and the south-east region, and so the problems of uneven and ineffective development, demonstrated under the LDDC, are likely to be repeated on an even larger scale.

The Thames Gateway provides new opportunities for the planning and design professions to engage in a more effective way with the rapidly changing process of urban development, and to learn from the experience of the past fourteen years in Docklands. The collaboration between government and large-scale private capital, and changing attitudes to planning, were powerfully demonstrated in the approach to Docklands redevelopment. There are many lessons to be drawn from this. If a better debate develops about the interlocking forces of urban change and effective strategies and techniques for planning, the next phase of London's development has a chance of being more successful for all concerned: citizens, government, developers and investors, and the planning and urban design professions.

NOTES

1. In a phased withdrawal, the LDDC will be reduced in size and will give up areas of Docklands in the period up to 31 March 1998. English Partnerships (a newly created regeneration agency) will take over parts of Docklands. Other areas will be given over to the Dockland's local authorities (e.g. Bermondsey Riverside area to Southwark Council in October 1994). The Royals or eastern part of Docklands are the largest undeveloped area to be given up by the LDDC.
2. SPLASH (South Poplar and Limehouse Action for Secure Housing) is a consortium of seven tenants' associations formed in 1990. As well as the legal action it is developing its own local plan. See SPLASH (1992) (available from Docklands Forum).
3. Negative equity exists where house prices fall below the level of housing loans, thereby trapping borrowers with loans that are higher than the potential returns from the sale of houses.
4. Estimates of the ratio of public to private expenditure vary. See Brownill (1990) and the Docklands Consultative Committee and LDDC references in the Bibliography.
5. A National Audit Office Report from the Department of Environment (HMSO, 1980), *Urban Development Corporations*, shows that Olympia and York paid £400,000 per acre for land when the market price was £3 million per acre. Olympia and York also agreed to contribute £150 million towards the Docklands Light Railway, and was committed to hiring 500 local construction workers; placing £2 million in trust for local schools; and supporting the river-bus between Charing Cross and Docklands. The Docklands Consultative Committee calculates that if Olympia and York had built its approved 12 million ft^2 it would have achieved a subsidy of £1.33 billion in deferred capital allowances (see also DCC 1990).
6. In its Corporate Plan of 1989, the LDDC claimed that it was currently 'devising a strategy that will enable all individuals and groups in London's Docklands to have equal access to the benefits and opportunities provided by regeneration' (LDDC 1989a: 5). The response of the indigenous population was 'better late than never', but optimism was dulled in that the LDDC cut its 'community division' budget from £43 million to £27 million in 1991, and again to £15 million in 1992.

BIBLIOGRAPHY

Ambrose, P. (1986) *Whatever Happened to Planning*, London: Methuen.

Anderson, J. (1990) 'The "New Right", enterprise zones and urban development corporations', *International Journal of Urban and Regional Research* 9(3): 468–89.

Attwood, C. (1989) 'London Docklands: urban design in regeneration', in A. Cortesi (ed.) *The City Tomorrow*, Florence: Aliena: pp. 121–43.

Barlow, J. and Gann, D. (1994) *Offices into Flats*, York: Joseph Rowntree Foundation.

Barrick, A. (1991) 'Docklands: an urban failure', *Building Design* (5 April): 1.

Breen, A. and Rigby, D. (1994) *Waterfronts: Cities Reclaim Their Edge*, New York: McGraw-Hill.

Brownill, S. (1990) *Developing London's Docklands: Another Great Planning Disaster?*, London: Paul Chapman.

Bruttomesso, R. (ed.) (1993) *Waterfronts: A New Frontier for Cities on Water*, Venice: Città d'Acqua.

Buchanan, P. (1989) 'Quays to design', *The Architectural Review* 1106: 39–44.

Budd, L. and Whimster, S. (1992) *Global Finance and Urban Living*, London: Routledge.

Cassidy, M. (1992) 'City changes: architecture in the City of London 1985–1995', London: Corporation of London.

Church, A. (1988) 'Demand-led planning, the inner-city crisis and the labour market: London Docklands evaluated', in B. Hoyle, D. Pinder and M. Husain (eds) *Revitalising the Waterfront: International Dimensions of Dockland Redevelopment*, London: Belhaven Press.

Coupland, A. (1988) *Housing in Docklands*, London: Joint Docklands Action Group.

Cullingworth, J. B. and Nadin, V. (1994) *Town and Country Planning in Britain*, London: Routledge.

Daniels, P. W. and Bobe, J. M. (1993) 'Extending the boundary of the City of London? The development of Canary Wharf', *Environment and Planning A* 25: 539–52.

Davies, C. (1987) 'Ad hoc in the docks', *The Architectural Review* 1080: 31–7.

Disraeli, B. (1881) *Sybil, or the Two Nations, By the Earl of Beaconsfield*, London: Longmans, Green.

Docklands Consultative Committee (1990) *The Docklands Experiment: A Critical Review of Eight Years of the LDDC*, London: DCC.

—— (1991) *10 Years of Docklands: How The Cake Was Cut*, London: Association of Metropolitan Authorities/DCC.

—— (1992) *All That Glitters Is Not Gold: A Critical Assessment of Canary Wharf*, London: DCC.

Docklands Forum (1990) *Employment in Docklands*, London: Docklands Forum, with Birkbeck College, London.
Edwards, B. (1992) *London's Docklands: Urban Design in an Age of Deregulation*, Oxford: Butterworth.
—— (1993) 'Deconstructing the City: the experience of London Docklands', *The Planner* (February): 16–18.
Fainstein, S. (1991) 'Promoting economic development', *American Planning Association Journal* 57(1): 22–33.
Glancey, J. (1991) 'Gotham City, E14', *The Independent* (13 July): 27.
Gosling, D. (1989) 'Urban design as public realm planning', in A. Cortesi (ed.) *The City Tomorrow*, Florence: Aliena: pp. 115–43.
Hall, P. (1992) 'Learning lessons from docklands', *Planning in London* (September) 3: 10–11.
Harvey, D. (1990) *The Condition of Postmodernity*, Oxford: Basil Blackwell.
Hatton, B. (1990) 'The development of London's Docklands', *Lotus International* 67: 55–89.
Hinsley, H. (1991) 'London's Docklands: the chance of a lifetime?', in R. Bruttomesso (ed.) *Waterfronts: A New Frontier for Cities on Water*, Venice: Città d'Acqua.
—— (1995a) 'Public/private interests in the redevelopment of London's Docklands', in A. Vernez Moudon (ed.) *Urban Design: Reshaping our Cities*, Seattle: University of Washington Press,
—— (1995b) 'Docklands is not the only future: some aspects of the redevelopment of London', in J. Walter, H. Hinsley and P. Spearritt (eds) *Changing Cities: Reflections on Britain and Australia*, Brisbane: Monash University Press.
HMSO (1984) *Advice on the Enterprise Zone Planning System in England and Wales*, London: Department of the Environment.
—— (1988) *Department of Environment: Urban Development Corporations*, London: National Audit Office.
Hollamby, E. (1990) *Docklands: London's Backyard into Front Yard*, London: Docklands Forum.
King, A. (1990) *Global Cities: Post-Imperialism and the Internationalisation of London*, London: Routledge.
London Borough of Southwark (1989) *Broken Promises: The Southwark Experience of the LDDC*, London: London Borough of Southwark.
London Docklands Development Corporation (1982) *Isle of Dogs: A Guide to Design and Development*, London: LDDC.
—— (1989) *Corporate Plan 1989*, London: LDDC.
—— (1982–95a) *The Corporate Plan*, London: LDDC.
—— (1982–95b) *Annual Report*, London: LDDC.
—— (1990a) *Review of Achievement*, London: LDDC.
—— (1990b) *London Docklands Architectural Review*, London: LDDC.
London Planning Advisory Committee (1991) *London: World City*, London: HMSO (report of research project co-ordinated by LPAC).
McIntosh, A. (1991) 'Another point of view on London's Docklands', in R. Bruttomesso (ed.) *Waterfronts: A New Frontier for Cities on Water*, Venice: Città d'Acqua: pp. 47–51.
Massey, D. (1991) *Docklands: A Microcosm of Broader Social and Economic Trends*, London: Docklands Forum.
Merrifield, A. (1993) 'The Canary Wharf debacle', *Environment and Planning A* 25: 1247–65.
National Audit Office (1988) *Department of the Environment Urban Development Corporations*, London: HMSO.

Sennett, R. (1990) *The Conscience of the Eye*, London: Faber and Faber.
Shaw, B. (1991) 'The London Docklands experience', in R. Bruttomesso (ed.) *Waterfronts: A New Frontier for Cities on Water*, Venice: Città d'Acqua: pp. 42–6.
Soja, E. (1988) *Postmodern Geographies: The Reassertion of Space in Critical Social Theory*, London: Verso.
Sorenson, R. (1993) *Building Design* (6 August): 4.
SPLASH (1992) 'The other side of Docklands' (available from Docklands Forum).
Thornley, A. (1991) *Urban Planning Under Thatcherism*, London: Routledge.
—— (ed.) (1992) *The Crisis of London*, London: Routledge.
Williams, S. (1990) *Docklands*, London: Architecture Design and Technology Press.

4

DUBLIN: MOTIVE, IMAGE AND REALITY IN THE CUSTOM HOUSE DOCKS

Patrick Malone

INTRODUCTION

One of the central questions in urban development is how far it is shaped by good practice in planning and design, rather than by the ambitions of property and political interests which hold power in the development process. The development of the Custom House Docks in Dublin (Ireland) provides an opportunity to examine how the objectives of property and other interests are integrated with planning and design processes. This raises the issue of whether the Custom House Docks project was created within a system of urban development that was based essentially on planning, or within a system in which planning and design were hijacked by property and other interests working for their own ends. In short, the issue is that of the motives behind the project and the position of political and economic interests and of planning and design in relation to those motives.

Officially, the Custom House Docks project was initiated in October 1985, when the Irish government decided to establish a development agency to revamp an area of redundant dockland and to use tax incentives to encourage investment in the area. However, the way in which the project was initiated and the nature of the interests behind it were generally hidden under a history of enabling legislation and planning documentation. The motives that formed the agenda for the project were masked by legislative and planning measures through which the project was merely 'institutionalized'. In reality, the Custom House Docks project was instigated, not by 'planning', but by a combination of political, property and other interests. Thus it is only by breaking into the relatively secret negotiations between those interests, and by exposing their often unstated ambitions, that the real purposes and meaning of the Custom House Docks may be understood.

Figure 4.1 Dublin: view of Custom House Docks, 1991
Source: Custom House Docks Development Company

THE OBJECTIVES OF THE CUSTOM HOUSE DOCKS

In 1986, the site of the Custom House Docks comprised roughly 11 ha of redundant dockland, situated at the boundary of Dublin's commercial core (Fig. 4.1). In 1993, when the development was scheduled to be finished, approximately one-third had been completed and the project was floundering.[1] In this respect the Custom House Docks is typical of many projects which were born in the euphoria of the mid-1980s, only to falter towards the end of the decade. The project was launched in 1986, alongside government efforts to promote urban renewal in a number of Irish cities. In the spirit of the 1980s, the Irish government laid down financial incentives to lure the private sector into urban renewal and to expand output and employment within a construction industry suffering the effects of recession. Thus, the Custom House Docks shares its legislative basis with a general programme for state-sponsored urban renewal in Dublin which is aimed at promoting development in neglected, secondary inner-city areas. This programme is also intended to draw developers and occupiers away from the prime areas of the commercial core, where the drive for relatively higher development profits has damaged the historical fabric of Georgian and Victorian Dublin (McDonald 1985).

The Custom House Docks can be distinguished from other designated areas in Dublin in a number of respects. Although relatively small in international terms, it was heralded as Dublin's largest commercial inner-city development and the most significant partnership of private and public interests in urban renewal in the history of the Irish Republic. It is unique in that it is under the management of a specially formed Custom House Docks Development Authority (CHDDA), which is exempt from local authority planning control and is empowered to acquire land, initiate and manage development, and dispose of completed office space. The project is also unique in terms of the financial incentives used to support it.

The incentives created to fund the general designated-areas policy in Dublin operate indirectly to boost rents, building prices and development profits by allowing occupiers relief from rates (property tax) and from income tax, in line with expenditure on buildings or rent. Within the Custom House Docks, the standard package of incentives is more generous than in other designated areas in Dublin. Moreover, the project is backed by additional tax incentives and concessions which were created to support the development of an International Financial Services Centre (IFSC) within the office element of the project.[2] As initially envisaged, the IFSC was to be substantially developed as a financial services centre for foreign and Irish companies conducting foreign business in non-Irish currencies. The primary IFSC incentive allows approved companies to pay only 10 per cent corporation tax on foreign business until 2006.

Whereas it is common practice in Ireland to use tax incentives to attract foreign investment into the industrial base of the economy, the IFSC represents an attempt by the government to draw international financial capital into Ireland. Thus the IFSC is important to the development of the Custom House Docks and to the growth of the financial services sector of the Irish economy. This means, however, that any assessment of the costs or benefits of the tax incentives used to support the IFSC must take into account its value in terms of urban renewal and its impact on the national economy.

The introduction of the IFSC into the Custom House Docks bound the state to the development at three levels. First, the state is involved through the development process in that it has financed the acquisition of land and the formation and operation of the development authority (CHDDA). Second, the state sponsors the package of financial incentives awarded to the project under the general designated-areas policy. Third, it provides the additional incentives used to support the IFSC. In the development of the IFSC and the Custom House Docks, emphasis has been laid on the principle of *leverage*. The state's input, which is mainly in the form of financial incentives, is expected to generate direct and indirect multiplier effects in terms of urban renewal, economic growth and increased employment.

In 1988, some fourteen months after its formation, the CHDDA made a development agreement with a consortium comprising one British and two Irish developers (British Land, Hardwicke and McInerney).[3] The project was to be completed in five years (by spring 1993) and was to consist of 150,000 m^2 of space, roughly half of which was to be office space for the IFSC.[4] In addition, the project was to contain a small marina, flanked by retail and leisure functions, pubs, restaurants, a cinema, museums and a hotel, with luxury housing set around two small dock basins. The retail and housing elements were to account for 8.5 and 13 per cent respectively of the total area of the project (CHDDA 1987: 11) (Figs 4.2 and 4.3).

The CHDDA aimed to develop an area where people would want to live and work. In addition, the area was to be enlivened by recreational and cultural activities. This was to be a 'centre of excellence and innovation' in urban regeneration. The brief prepared by the CHDDA to elicit proposals from developers, and the scheme put forward by the chosen development consortium, provided a battery of images and intentions suggesting a lively, attractive, mixed-use development that would be active outside office hours. Importance was given to housing as part of a 'living development', to leisure and cultural functions and to the exploitation of waterfront amenities. The integration of the development with the existing commercial core, and the flow of people to and from the core, were also important. It was intended to convert a redundant area of dockland into a flagship development situated within a duly enlarged core. Moreover, the project was to act as a bridge that would draw life from the core while stimulating further development within a wider area of the docklands. It was to enhance the city and, at the local level, to benefit one of the poorest areas of Dublin in the form of jobs, amenities and a better living environment (CHDDA 1988: 24).

THE REALITY OF THE CUSTOM HOUSE DOCKS

One year after the proposed date of completion, only 40 per cent of the project has been realized. The development as it now stands (in October 1995) consists almost entirely of office space.[5] There is no sign of the marina, retail and leisure elements (other than one pub). While the hotel is under way, the cinema and the three museums have not been constructed yet.

Because the office component of the project has faced difficulties in terms of demand and the growth of the IFSC, the developers have been reluctant to press ahead with further office development and, consequently, with the other, less profitable elements of the project. For the developers, disagreements have centred on the nature of the office content of the project, on profitability and the funding of the less lucrative parts of the development. Publicly, the development has been drenched in positive

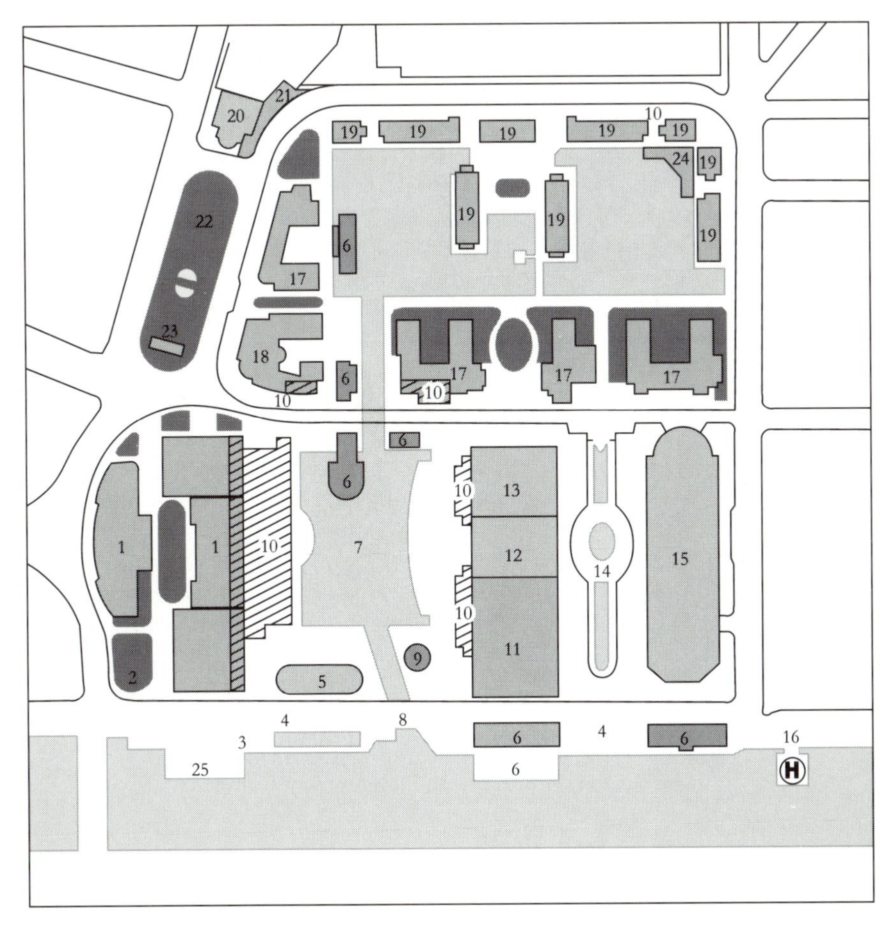

KEY SERVICES

1. International Financial Services Centre
2. Fountain Plaza
3. Open-air Market
4. Riverside Park
5. Museum of Folk Art
6. Restaurants and Bars
7. George's Dock Marina
8. Bridge
9. Bandstand Pavilion
10. Retail
11. Museum of Modern Art
    Museum of Science
    Flea Market & Antique Shops
    Lecture Theatre
12. Winter Garden
13. Cinemas and Nightclubs
14. Reflecting Pool and Sculpture Garden
15. Hotel
16. Heliport
17. Financial Services Office Buildings
18. International Trade Centre/ Financial Services
19. Residential
20. Connolly Station
21. Community Centre
22. Station Park
23. The Original Stone Gateway
24. Boat Club
25. Canoe Club

Figure 4.2 Dublin: plan of Custom House Docks, 1988
Source: Custom House Docks Development Company

imagery. Privately, a network of tensions and disagreements has emerged between the government, the CHDDA and the development consortium. The development came to a virtual standstill in the early 1990s. However,

Figure 4.3 Dublin: model of Custom House Docks, 1988
Source: Custom House Docks Development Company

the mid-1990s have seen the addition of a relatively small office block and the housing element of the project.

The fact that some building work recommenced after a period of relative inertia may reflect the revised conditions set out in a new agreement with the development consortium and factors associated with the incentives used to support the project.[6] Moreover, the CHDDA has taken a more assertive role in terms of the development process and the expansion of the project into adjoining areas. However, the initial proposals for the project and the first development agreement have been retarded by lack of demand for office space in the IFSC and by development profits which have been insufficient to fuel the construction of the less profitable parts of the development.

Insomuch as the economics of the Custom House Docks project hangs on the success of the IFSC, it should be noted that the IFSC was hatched at a time of great optimism in the financial markets. It represented Ireland's bid to tap into flows of international financial capital. However, the project had to compete with other international financial centres for a limited volume of capital. In order to attract occupiers into the IFSC, the range of IFSC tax incentives has been extended. Special incentives have been awarded to particular sectors of the financial services industry and to IFSC employees. In addition, the range of activities considered appropriate for IFSC benefits has been stretched to cover allied 'back office' operations such as data processing, information services and share registration. The development has also been facilitated by a flexible attitude to the terms under which leases may be held. However, despite the initial emphasis on its potential to attract foreign economic activity, the IFSC as it now stands is occupied predominantly by Irish firms. Some companies conduct foreign business in foreign currencies as part of their operations on the site, others have no foreign business in the terms of the IFSC. Moreover, many companies were operating elsewhere in Dublin prior to moving into the Custom House Docks, prompting critics to suggest that existing companies, notably the major Irish banks, are being heavily sponsored simply to move offices within Dublin. It is also important to note that firms approved for entry into the IFSC, and supposedly waiting to move to the Custom House Docks, may benefit from the 10 per cent corporation tax concession while operating outside the Custom House Docks site. By the middle of 1994, 260 firms had been approved for entry into the IFSC, of which 180 were in operation and benefiting from 10 per cent corporation tax and other IFSC incentives. However, only 40 (22 per cent) of the firms operating under the IFSC umbrella were actually located within the Custom House Docks.

Many IFSC firms may be reluctant to move into the Custom House Docks given that they already benefit from the most significant tax concession associated with the project. However, the failure (or reluctance) of the government to force IFSC companies into the Custom House Docks has clearly weakened the link between the IFSC and the Custom House Docks and lowered the value of the IFSC as a supportive element of the project. Moreover, the decision to allow companies to benefit from the IFSC concessions 'prior to entry' into the Custom House Docks can only be interpreted as a blunder, or as a loophole through which the IFSC benefits can be spread beyond the Custom House Docks site.

The CHDDA, anxious to maintain the objectives of the IFSC, has attempted to insist that the Irish interests that now control the existing office space within the Custom House Docks should let any further space to interests (Irish or foreign) conducting foreign business under the terms of the IFSC. However, efforts to let space, or to pre-let proposed new office space, have been frustrated by weak demand and by the failure of the government to force approved IFSC companies, operating outside the site, into the Custom House Docks.

The demand for office space may also be frustrated by rents that are roughly twice those of prime space outside the project, although this problem is temporarily offset by the designated-areas tax incentive which allows occupiers a double rent allowance against tax for ten years. Demand is also restricted in that many firms want to rent 500 m^2 or less – a factor that adds to the impression that the IFSC may attract occupiers who set up small front offices to launder foreign-currency operations within a partial tax haven.

Many IFSC companies located outside the Custom House Docks want, as a condition of entry to the site, to break free of their leases when the tax concessions expire in 2006. This raises the question of what will happen within the IFSC and the Custom House Docks when the financial incentives associated with the designated-areas policy and the IFSC are withdrawn. The IFSC corporation tax incentive, which has been extended once, is due to expire in terms of new applications at the end of 1994. The double rent allowance, which has been extended twice, will be removed at the end of 1997. While the expiry of the designated-area incentives may have some effect on rent levels, the expiry of the IFSC corporation tax concession will raise the crucial question facing governments that use tax incentives to attract foreign companies: do the companies remain after the tax concessions have run out?

The IFSC has also exposed the dangers presented by foreign governments that attempt to counteract the effects of foreign tax havens on their economies. Used as bait to catch foreign companies, the IFSC incentives are intended to draw economic activity into Ireland that might be subject to higher taxation in another country. For example, a German company opting into the IFSC may establish a small 'front office' in Dublin to 'launder' its pensions business, thereby avoiding substantial German taxes at the comparably small cost of 10 per cent corporation tax in Ireland. In 1991, the German government's reactions to the threat posed by the IFSC demonstrated that the capacity of the IFSC to draw companies into Ireland depends on two factors which must be finely balanced: the generosity of Irish tax concessions relative to concessions offered elsewhere, and the tolerance of foreign governments which stand to lose revenue to the Irish exchequer.[7]

In summary, it can be said that the Custom House Docks has been beset by various difficulties. Although ostensibly a faltering exercise in urban renewal, the project is driven essentially by economic forces and hindered by economic problems. This suggests that an examination of the economics of the project might expose the primary motives behind it, and provide a basis on which it might be interpreted and assessed.

INTERPRETING THE CUSTOM HOUSE DOCKS

It is remarkably difficult to know precisely why the Custom House Docks project exists. Its promoters have invested in marketing that suggests how the project was initiated and that details the purposes of the project in terms of the city, the economy and the local community. However, this only serves to reinforce the impression that there is a hidden agenda set by the major interests behind the project, and that the meaning of the project can only be established by exposing those interests and their motives. Alternatively, it is possible to adopt the potentially naïve assumption that the project has been driven by objectives that relate to the national economy and to urban renewal, and that the purposes of the project might be exposed by examining its logic in economic and planning terms. If, in adopting this assumption, however, it is not possible to validate the project, then an even greater emphasis must be placed on the motives of the interests behind its development.

In economic terms, it is difficult to undertake a detailed assessment of the Custom House Docks. The project has been incorrectly described as involving no public expenditure, on the basis that the money provided by the state to purchase land and run the CHDDA may be returned through premiums paid by the developers and the state's share of development profits. If we ignore any infrastructure costs borne by local authorities, it may be that the state will recover the public money invested directly in the development process, in land and the CHDDA. However, the state-sponsored incentives granted to occupiers are more significant than the state's sponsorship of the development process. Thus, in calculating the state's investment in the project, account must be taken of the cost of the designated-area and IFSC incentives, and especially the cost of the IFSC's 10 per cent corporation tax incentive.

Apologists for incentives might argue that tax-led development policies should be assessed on the basis of their effectiveness. In other words, the validity of the state's sponsorship of private interests might be tested by assessing the costs and benefits of incentives and tax breaks. In the case of the Custom House Docks and the IFSC, however, rigorous testing may be impossible. One reason for this is that the project would need to be assessed fully on two interrelated fronts: as an exercise in urban renewal and in terms of the IFSC's impact on the Irish economy. The designated-area and IFSC incentives have together contributed to the physical and economic aspects of the project, although the fact that many IFSC firms are located outside the Custom House Docks causes some divergence between the growth of the IFSC and the development of the Custom House Docks.

If the reciprocity between the Custom House Docks and the IFSC is ignored, the basic question is whether the cost of the IFSC incentives is justified by social and economic gains produced anywhere within the economic system. If tax incentives have been awarded to firms on the basis of economic activity that would have existed regardless of these incentives, then the use of incentives is clearly questionable. In short, to gain 10 per cent tax from economic activity that would not have existed otherwise is positive, but to lose 30 per cent of the tax on economic activity that would have existed regardless of tax incentives is clearly negative. A similar point may be made with respect to the impact of incentives on employment. It is said that the IFSC has 'created' 1,400 jobs, but this is not to suggest that

IFSC incentives have generated a net increase in employment of 1,400 jobs (Government of Ireland 1994a: 23).[8]

Although the validity of any IFSC incentive hinges on how much new economic activity it generates, this is difficult if not impossible to establish. A full assessment would require knowledge of the complete range of direct and indirect economic and social benefits generated by an incentive. It is difficult even to know at what point in the 'history' of an incentive its benefits and disadvantages might be assessed fairly. There is also the difficulty of assessing the impact of any individual incentive when companies and the economy are responding to a package of incentives. Moreover, the significance of incentives must be measured against all other factors that determine location.

Although the costs and benefits of the incentives used to support the IFSC and the Custom House Docks may be difficult to assess, it is clear that they are significant in terms of lost tax revenue. The capital cost of the Custom House Docks project, estimated in terms of the cost of buildings to investors, can be written off against tax. Leaseholders may deduct from tax twice the value of rents paid over ten years, in a project where rents are roughly twice those of prime office space in Dublin. The cost to the state of the remission of rates (property tax) may be relatively small. In contrast, the net cost of the IFSC corporation tax incentive is potentially high, particularly if one takes a cynical view of the volume of new economic activity that may result from that incentive.

The tax revenue from the IFSC after the year 2006 is predicted to be in the region of IR£400 million per year. If it is allowed that the gross average taxable income generated annually by IFSC companies which enjoy the 10 per cent tax concession is only half of this amount (IR£200 million), this would mean an average tax write-off of roughly IR£150 million a year, assuming that those companies would normally be taxed at 40 per cent.[9] On this basis, over a ten-year period the cost of the 10 per cent corporation tax incentive might be approximately four times greater than the total capital cost of the Custom House Docks project.[10]

It is important to note, however, that the overall cost of the state's sponsorship of the IFSC and the Custom House Docks must be assessed against their role in relation to the long-term prospects for growth in the financial sector. Whereas both the costs and the efficacy of incentives may be difficult to measure, interpretations of the IFSC and its prospects vary. For example, some apologists are keen to stress the success of the project in terms of the flow of funds into Dublin in the mid-1990s. However, a recent North American review of the IFSC demonstrates the dilemma for governments which must attract financial capital in a competitive global economy in that:

> IFSC collective investment funds are exempt from tax on income or

> capital gains and free from withholding taxes on dividend distributions. Ireland also has the lowest corporate tax rate in Europe: a 10 percent tax limit on the net income of fund management companies. Overseas investors, moreover, are exempt from stamp duty, gift or inheritance tax and value-added tax. Ireland's high unemployment rate means that there are hundreds of extremely well-educated applicants for every IFSC job advertised . . . who are willing work for lower salaries. They are also willing to adopt the American work culture of long days and sparse holidays, unimaginable in much of Europe.
>
> (Sullivan 1995)

The planning perspective

Given that it is difficult to validate the Custom House Docks on the basis of broad economic and social effects, it is also difficult to establish the validity of the project, and of the state's sponsorship of the project, on the basis of its implications for planning and urban renewal. It could be argued, for example, that the project has been given preferential treatment, while other inner-city areas are in greater need of redevelopment, or might have been redeveloped with greater benefit. Much as politicians may champion the cause of dockland areas, these are developed at the expense of other parts of the city. This raises issues concerning the management of development pressure and the optimization of development activity within the urban structure. It has been suggested that the process of designating areas and distributing incentives in Dublin 'has followed no scientific basis' either in terms of deciding which areas should be designated or in assessing the effects on areas which are marginalized by the designated-area policy (Corrigan n.d.: 26).

The office element of the Custom House Docks raises similar questions, in that there is evidence that the project was launched with little regard for the overall market for office space in Dublin (Malone 1990: 9). Although the IFSC was intended to accommodate predominantly new and imported office functions, it has drawn into the Custom House Docks existing and perhaps potential occupiers from the general market for office space. It could be argued that this depressed the general or 'natural' property markets, as rents, capital values and land prices adjusted to compete with the state-sponsored IFSC. Paradoxically, however, a short and relatively artificial 'boom' in the office development cycle in Dublin may have been partly stimulated by the initial spirit of optimism surrounding the Custom House Docks. Thus, the project may have stimulated office development at a general level, while having the potential to depress markets by producing enough office space to soak up the equivalent of two or three years' demand in the Dublin office market (Malone 1990: 35).

One of the major issues raised by the incentives is the extent to which a state-sponsored project may simply involve meeting demand from the general market at a particular location, as opposed to a situation where development soaks up new or specialist areas of demand created by incentives. Moreover, in that demand, development pressure and the level of activity in the construction industry are interrelated, any claim that the Custom House Docks has boosted the overall level of activity in the construction industry can only be justified on the basis that it has increased (through economic effects) the general level of demand for new space. If the project has simply siphoned off building activity or the demand for space into a privileged development supported by state incentives, then the use of incentives must be open to question. The Irish Minister for Finance has said that 'the great thing is that urban renewal is labour intensive', the implication being that it may be used, and sponsored by the government, to soak up unemployment and under-utilization in the building industry (Ahern 1994: 6). Yet the government also recognizes that 'in the medium to long term, construction investment cannot grow faster than the overall rate of growth in the economy' and that, at best, urban development can only provide 'the essential infrastructure for further economic expansion' (Government of Ireland 1994a: 18). In other words, whereas construction can provide a platform for economic growth, development must be underpinned by economic growth and the demand for new space. In this respect, the use of incentives in urban renewal in Ireland may already have disturbed the relationship between output in the construction industry and the level of demand generated by economic growth (DoE n.d.: 27).

In its current state, the Custom House Docks can be seen as a faltering exercise in urban renewal that has absorbed, and will continue to absorb, a large measure of public money. Although the validity of the state's sponsorship of the project is difficult to test in economic, social or planning terms, there is a sense that the project is the product of a crude and opportunistic capitalism that has more to do with private ambitions than with good planning or public needs. Whatever the objectives listed in support of incentives, the state's sponsorship of the IFSC and the Custom House Docks fuels the ambitions of property, political and financial interests (Benson 1991: 5; Kavanagh 1991: 102). It augments the profits of the traditional landed, development and property investment interests, and spawns new opportunities for profit, for example, for estate agents and financial advisers. In this respect, the use of incentives can modify the traditional structure of relations between the interests involved with urban development and the flows of profit and financial benefits generated by development. The incentives used to support the Custom House Docks demonstrate how occupiers now benefit from planning policies, and how the financial benefits associated with development may be spread beyond the traditional network of property and investment interests. It would be interesting, for example, to compare the financial value of incentives enjoyed by the major banks in the IFSC with the profits drawn by the development consortium.[11] Essentially, however, the question is how far the use of public money to augment profits is valid, and whether state-sponsorship was introduced to secure urban planning or social aims or to feed the narrow ambitions of financial, property or political interests. This question is more acute where the economic, social and planning benefits of sponsorship are in doubt.

One criticism which might be levelled at the Custom House Docks can be directed at urban development in Dublin as a whole: namely, that development is orientated primarily towards lining the pockets of a raft of property, financial and other interests. Urban development in Dublin is rooted in the drive for profit. Social objectives, and the wider purposes of good planning, are generally subservient. Thus, there is an emphasis on incentives and incentive-led development, rather than on the power of planning to restrict or control development processes. As befits a predominantly profit-orientated system of development, the stick gives way to the carrot. Arguably, there is a social component to planning in Dublin, but this is relatively weak in a system dominated by the market and private capital. Given the nature of planning in Dublin, it could be said that wider social and planning objectives can only exist within the slipstream of development processes driven by the urge for profit. The manipulation of state-sponsored incentives may offer some opportunities for planning to lead the flow of development capital into secondary city areas, or into neglected forms of development such as housing or the rehabilitation of existing buildings. However, planning is left to play a secondary and facilitating role in a system that is geared to the generation of various forms of profit for the different interests associated with urban development.

Given that the tax incentives used to support the Custom House Docks and the IFSC may be expensive and wasteful of state resources, it may be fortunate for the government that it is difficult, if not impossible, to assess the validity of these incentives. Government and other interests may be protected from criticism by the difficulty of analysing the costs of incentives, and by the secrecy with which state and private interests treat the limited information on which the value of incentives might be assessed. For the same reasons, it is clear that any claims made by politicians or others regarding the benefits of incentives are likely to be ideological or opportunistic rather than scientific.

SELLING THE CUSTOM HOUSE DOCKS: THE ROLE OF PLANNING AND DESIGN

As it stands, the Custom House Docks project does not measure up to the physical, social or economic objectives set at its inception. The development is unlikely to substantiate the imagery that smoothed its launch. Two key objectives have been elusive: the construction of a vibrant mixed-use development and the creation (within the development) of a large financial services centre for new and foreign firms. Although the project is substantially incomplete, what the public now see is a relatively large office development with a small area of private housing. There is no hint of the 'people place', of the intensively occupied mixed-use development with a festive atmosphere (Fig. 4.4). Instead, there is cold granite, green glass and an air of distant bureaucracy; in short, what passes in Ireland for bankers' architecture (Fig. 4.5).

It could be argued that the difficulties faced by the project are due to bad planning, as demonstrated, for example, by the failure to predict the problem of insufficient demand. Given that planning should direct urban development to the best social advantage, it is difficult to be convinced that the Custom House Docks is the product of good planning. Major planning problems, which eventually impeded the project, were not fully explored or articulated in the initial planning and design phases. For example, more attention might have been paid to the difficulties of promoting retail and leisure functions on a site that is tangential to the commercial core and isolated by traffic. However, to throw doubt on the Custom House Docks as a product of planning may serve only to obscure a more fundamental issue. The key question is not whether the project has been the subject of errors in planning, but rather whether it is rooted primarily in planning.

Arguably, the Custom House Docks project stems from a system of development that is not fundamentally led by planning, but by property and other interests that hijacked development processes for their own ends. In this system, the planning and design mechanisms associated with the project were essentially geared to generating a stream of positive imagery, and were partly subsumed into the marketing process that launched the project. This prompts the suggestion that planning and design, while geared to the production of imagery, may have actually frustrated the pursuit of good planning practice and obscured rather than illuminated the deeper realities of the project. Moreover, planning and design may also have helped to conceal the real intentions of the major interests behind the project, which in retrospect appear to have been buried under a superstructure of hopes, promises and marketing imagery depicting the potential benefits of the project for dockland areas, the city, the community and the national economy.

The marketing role of planning and design may have been facilitated by the CHDDA's emphasis on 'deliverability', and on the speed with which the project moved from initiation to the appointment of the development consortium. In that this happened within a period of roughly one year, there was perhaps insufficient time to resolve some of the issues raised by the development. The Custom House Docks may also serve to demonstrate that the use of design competitions for relatively large projects can have

Figure 4.4 Dublin: views of proposed Custom House Docks, 1988
Source: Custom House Docks Development Company

Figure 4.5 Dublin: anniversary celebrations by IFSC employees, 1993
Source: Tony O'Shea, *The Sunday Business Post*, 22 August 1993

negative effects. Those setting the brief may be required to put together a project in a short time, and competitors may be forced to make relatively superficial responses based on inadequate analyses, which later provide weak cornerstones for further development.

In the 1980s, the use of architectural competitions reflected a general tendency towards image-making, marketing and media hype. This was the era of postmodernist 'commodification' and of a new coalition of aesthetics and capital (Harvey 1989). Marketing, and a tendency to decorate the truth, became an important part of the work of professionals and other interests involved in development. To front major development projects, the marketing process fabricated fantasies. Partly inspired by international practice, the redevelopment of the Custom House Docks was fronted by various levels of marketing aimed at different audiences. Each audience might be defined on the basis of its relationship to – and perhaps its geographical distance from – the project.

An interesting characteristic of marketing is how various images can be attached to what is essentially one development or one physical object. In terms of the Custom House Docks, the imagery surrounding the project falls into three categories. As discussed above, one layer of design and planning imagery was laid down in the initial phases of the project. A second set of imagery was coined to market the development to potential occupiers, and specifically to promote international interest in the IFSC. Third, at the local level, the relatively poor and 'socially residual' residents of the area around the Custom House Docks were treated to imagery that promised employment, job training and environmental improvements. For its Irish audience, the CHDDA sought to reform the perception of the area around the Custom House Docks, and to shift the emphasis from the area's problems to the opportunities afforded by redevelopment. A local community liaison policy was based on the four key features of employment, training, the encouragement of local business, and environmental improvements (CHDDA 1989: 24). In reality, few training opportunities were provided for local young people. Ultimately, the most discernible impact on the local population was probably the removal of residents from public housing adjoining the site, and the absorption of land previously occupied by that public housing into an enlarged Custom House Docks site.

At the international level, the CHDDA and the state's Industrial Development Authority (IDA) have promoted the IFSC principally in terms of the financial incentives offered to foreign investors. The incentives are gift-wrapped in imagery portraying Ireland in terms of political stability, an attractive lifestyle and a willing and educated work-force. Many Irish people, and particularly residents of the area around the Custom House Docks, might have difficulty recognizing the images offered to foreigners to lure them into the IFSC. An Ireland of expensively dressed university students, high-tech electronics, quality consumer goods, excellent medical facilities, golf clubs and rivers with abundant fish, is at odds with the poverty, crime and dereliction that many associate with inner-city Dublin.

The imagery associated with the Custom House Docks can also be classified according to its marketing function and intended audience. Whereas the imagery created around the IFSC reflects high technology, sophisticated communication systems, high finance and an élitist 'lifestyle', the architectural and design imagery surrounding the Custom House Docks falls into the 'festival tradition' in design and waterfront development. The buzz phrases are 'festival shopping', 'livability' and 'people place'. Drawn from the 'festive townscape' tradition in design, this imagery is reminiscent of waterfront developments in Boston and Baltimore, of the American developer Rouse, and the work of the architect, Benjamin Thompson, who played a major role in the launch of the Custom House Docks project. It is also reminiscent of the picturesque imagery of Gordon Cullen but, in that it is based on shallow intentions, the festival tradition is more akin to visual confectionery.

In examining the design imagery of the Custom House Docks, it must be remembered that the development was to be 'fronted', both physically and in terms of the construction process, by a major office development housing the IFSC. This was to be the 'flagship' of the development, located on the most public boundary of the site and intended to give an impression of high-quality 'state of the art' office space. To suggest 'livability', however, the imagery produced in the initial stages of the project overlaid the office content with images of pedestrianized areas with shops, restaurants, housing and cultural facilities. This offset the uniformity of the office buildings, while the images of 'festival shopping' and lively leisure activities suggested that the development would be active throughout the day. The housing element was labelled 'interactive' and was set 'in a neighbourly enclave which embraces its own urban lake'. Weight was given to the historic ambience of the site and to its existing buildings and artefacts. Moreover, the development was to generate a 'people place' along the quayside and exploit the waterfront for recreation and transportation. Essentially, the site and the adjoining quayside were to operate as a 'major focus for city life', to provide an integrated working, living and leisure environment in the form of an 'urban microcosm' and a 'vibrant people place'. Within the site, the imagery was of an unfolding pageant of events, 'a variety of sensory delights, with interesting merchandising, street activity and opportunities for lunching, dining and entertainment' (Benson 1991: 16–17).

It is interesting that the initial imagery coined by the CHDDA was compatible with that of the American architect Benjamin Thompson, who later became formally involved with the development.[12] In the period immediately after the 'competition' the architectural press, with its usual innocence, helped to portray the project in festive 'people place' imagery. For example, in the British journal, *Building Design*, the project was illustrated using text provided by the promoters of the project, which described life within the Custom House Docks at some time in the future:

> It is a warm, calm September evening. The highly-paid executives in the Financial Services Centre are still at work – their vdu's giving out the latest on Wall Street. At the Liffey's edge the tanned and fit members of the Custom House yacht club are tying up their craft and are strolling leisurely to the dockside pub for a pint or G&T. The kids are not yet back at school. . . . The culture vultures are on their third museum – in the Dublin section they are still not over the shock of what the city was like when it had vacant sites. At the Heliport a Ryan Air courtesy helicopter arrives with some more tourists. A limousine whisks them to their luxury hotel. In the apartments a successful young barrister has just arrived home via a vaporetto from the law courts up the quays. She sits on her penthouse balcony admiring the spectacular view of the mountains. As she sips her Campari soda she wonders if the Bunuel movie is playing at the Screen on the Dock . . .
>
> (Cahill 1988: 22)

It will come as no surprise that the Custom House Docks, as it now stands, reflects nothing of this imagery. Neither does anything in this imagery suggest the darker realities of the inner city. There is no hint here of crime, poverty, drug abuse or dereliction, although these problems surround the site of the Custom House Docks and might be visible from the balcony of the Campari-sipping barrister, should she shift her gaze from the mountains, which are far to the south, to examine areas closer to hand. However, the tunnel vision of the barrister reflects the fact that the festive, mixed-use formula adopted to launch the project afforded attractive marketing imagery that allowed the social conditions surrounding the site to be swept under the carpet. It allowed the office content of the development to be overlaid with associations of pleasure and 'culture'. In retrospect, however, it seems that the planning and design processes that generated this imagery had more to do with marketing than with the difficulties that would confront the project.

POWER, MOTIVE AND PROFIT

To establish the position of architecture, planning and urban design in the Custom House Docks, it is necessary to understand the primary motives behind the project, and to know how, why and by whom it was conceived. It short, it is necessary to discern how urban design and planning have been interwoven with a network of economic, political and other goals. By understanding the central motives and specifically human origins of the project, it may be possible to uncover the relationships between those motives, the objectives and imagery generated by planning and design, and the reality of the project as it exists. It is difficult, however, to expose the web of interests behind the project. In theory, it might be possible to identify the government, development, property and other interests associated with the project; to expose the motives which they have pursued; and to measure the relative degree of power wielded by each interest, or each band of compatible interests. It might also be possible to extend this analysis into the secondary layer of bureaucrats, agents and design interests associated with the project, some of whom may have entered the arena after its initiation. A fully developed analysis could expose various interests, with various degrees of power, pursuing different and even conflicting goals. It might demonstrate that there are relative winners and losers in an 'arena of power' and that the search for profit, although a primary ambition for many interests, exists alongside political, personal and even social objectives. Moreover, it might be possible to overlay this analysis with another showing the flows of profit and financial benefits arising from the project. This would expose the specific relationships between power, motive and profit, and would, incidentally, expose the role of financial incentives associated with the project.

To fully 'model' the forces that acted on the project would require unravelling the network of alliances between the primary interests. It would also require the identification of the relative significance of central and peripheral interests, local political forces, the Port and Docks Board, key figures in the CHDDA and the development consortium, and any interest that may have had an impact on the project. Together, these interests created a cluster of generating motives, the most powerful of which formed the backbone of the development. Of the medley of goals and ambitions that have shaped the project, it might be assumed that the main goals were established by those who held the greatest power and who could thus determine the 'genetics' of the development. However, power should not be confused with success, in that interests which have played a primary role in the project may have fallen short of their objectives.[13]

Although it is possible to sketch theoretical models, or to debate the general nature of motives, power and profit in urban development, exposing these factors for a specific development can be difficult. Research may be impeded by secrecy, by the sensitivity of the information that is sought and by a tendency to 'institutionalize' the history of the project in a way that obscures its real origins. For example, the official history of the Custom House Docks suggests that it was initiated by the government and managed by the CHDDA. By highlighting the role of 'institutions', however, this history conceals the human origins of the project. It begs the question of just what is meant by 'the government', when political, business and property interests are interwoven. Thus, in order to unravel the 'genetics' of the Custom House Docks, it is necessary to uncover the involvement of specific individuals, to identify their motives and to weigh accurately the relative power behind these motives. Whether these interests are labelled, for example, as politicians, bankers or property interests, they are more accurately identified on the basis of the power that they wield and the motives which they bring to the development process. The problem is, however, that this is not the stuff of official records, but of a fundamental layer of private traffic between key interests working primarily for their own ends and under conditions of relative secrecy.

Within the hidden layer of primary motives driving the Custom House Docks, the issue for planning and design is the extent to which the urge for good practice formed part of the backbone of the project. How far were planning and design hijacked to provide marketing imagery, or to mask the realities of the development process and the primary motives or economic functions of the development? At one extreme, it might be argued that the planning objectives and design imagery associated with the Custom House Docks were determined principally by the need to market the project and to generate an ideological and largely illusory context within which the real project was surreptitiously created. Architectural imagery in the 'festival' or 'people place' tradition is particularly palatable. It can provide good camouflage for development interests, while rendering a proposed project acceptable to the public. However, it may be too cynical to suggest that none of those involved with the Custom House Docks intended to create a vibrant, well-occupied, mixed-use development. The 'festival' imagery may have seduced some of the seducers, veiled intractable realities and shrouded the invalidity of claims regarding the planning and social benefits of the project. Some of the shortcomings of the Custom House Docks may stem from the fact that imagery became a substitute for a deeper analysis of the problems facing the development, notably perhaps the problem of demand for the space that the development might generate. The key question, however, is whether that imagery also hid, or was intended to hide, the real nature of the motives behind the project. Did planning and design contribute to the process of 'commodification', evolving images of pleasure, spectacle and social benefit to obscure the realities of the development both in terms of the motives behind it and the planning and design problems which confronted it?

The Custom House Docks was to be a prestigious flagship project. It was to enlarge the commercial core by creating a densely occupied, mixed-use development that would enhance the inner city and encourage further development within Dublin's docklands. Some might argue that, given time, the project will reach a more successful conclusion or that it might be restructured to counter some of the planning problems which have impeded it. But the fundamental issue is not whether the project represents good or bad planning, or whether it might be restructured to offset planning difficulties. The question is whether the project reflects deeper problems in a system of urban development which is based not on planning, but on the crude logistics of capitalism, a logic that is so crude that even the urge for profit that drove many of the project's major players has not been realized satisfactorily in the development process. In this respect, the project as it now stands may be a perfect expression of the logistics of the development system that created it and of the intentions of those who held power in the development process.

CONCLUSIONS

The development of the Custom House Docks prompts questions about the extent to which planning and urban design are autonomous elements in the development process, and how far planners, architects and urban designers are free to pursue whatever constitutes 'best practice'. The project suggests that planning and urban design may be subordinated to the marketing process, or shaped by political and profit motives that are not obviously positive in terms of their social, planning or economic effects. At one extreme, the Custom House Docks may be seen to have been initiated and shaped by narrow property, business and political interests, masked by seductive design and planning objectives. Clearly, profit is central, whether assessed in terms of rent, development profit, or the incentives absorbed by occupiers. But the fact that a project may be driven by ambitions for profit, or by a crude urge to exploit urban development for private gain, does not guarantee its success in those terms. Rather, it seems that coarse exercises in exploitation tend to be self-defeating. Just as planning is thrown into the back seat, so too is any tendency to test the feasibility of projects in terms of, for example, demand or difficulties presented by the site. Raw capitalism, whether in London Docklands or Custom House Docks, does not lead to good planning; nor does it result in the efficient exploitation of urban development processes by political, property and other economic interests.

There is of course an alternative view of the Custom House Docks: a view in which the goals of promoting economic and social gains and of regenerating an area of the city are seen as evidence of the good faith of the project's creators and managers. Perhaps the truth is somewhere between these two viewpoints, although it may alter towards one end of the continuum or the other depending on the set of interests or motives examined. Obviously, the various political, property, business, bureaucratic and design interests involved with the project may be distinguished according to their relative power in the development process and the nature of their ambitions. An important distinction could be drawn between the political, property and financial interests that hold the greatest power and that set the agenda for the project, and the bureaucrats, designers and managers who may (perhaps regretfully) collude with economic and political interests that regard planning and design as enabling tools rather than as instruments of wider social or planning goals.

NOTES

1. Adjoining parcels of land have been added to the original site. At the time of writing (May 1994) the area of the Custom House Dock site was roughly 50 acres (20 ha). Comments regarding completion refer to the initial project, for which the annual report of the CHDDA for 1987 shows a total gross floor area of 147,300 m^2, of which 70,600 is offices, 27,900 is a hotel, 12,500 is retail space, 19,000 is set aside for 200 apartments and 17,300 for museum, cultural, leisure and community facilities.
2. The financial incentives for the Custom House Docks under the designated-area policy are: 100 per cent capital allowances on investment in commercial buildings; double rent allowance for ten years; relief from rates; and 'section 23' allowances for rented residential accommodation. Adjustments have been made, since 1986, to extend the time period over which incentives are available. Other adjustments have been made with respect to incentives for housing. The principal concession for IFSC companies is the 10 per cent corporation tax allowance (extended to the year 2006); however, there are many other general IFSC concessions and specific concessions apply to specific types of funds.
3. Under this agreement, the development was to be funded by the developers, who were guaranteed a specified economic return. Any surplus development profit was to be shared between the developers and the state on a 60/40 per cent basis in favour of the state. In addition, the developers were to pay a lump-sum premium to the state. McInerney has since dropped out of the consortium, which now consists of British Land and Hardwicke.
4. The initial details given out by the CHDDA are vague with respect to the exact area of office space to be taken by the IFSC.
5. In 1993, when the development was to have been finished, six office blocks were completed, with a total net area of roughly 40,000 m^2. This represents approximately 70 per cent of the proposed area for offices in the initial programme and roughly 34 per cent of the total area of the initial project (given as 70,600 and 147,300 m^2 gross, respectively). At the time of writing, additional space had been completed, an additional office development was under construction and groundwork was under way for 200 apartments. The total built area of the project in spring 1994 might be taken as roughly 40 per cent of the initial total of the proposed project. At the time of writing, work has commenced on the housing component of the project, which has been increased recently from 200 to 333 apartments.
6. With regard to incentives, the current situation is complex in that a two-year extension (to 24 January 1999) has been granted to cover the double-rent allowance. However, lobbying is under way to gain an extension for the IFSC corporation tax incentive, which would extend the period for the granting of licences to firms entering the IFSC, to beyond December 1994. The fact that the power to grant licences is set to expire may encourage some immediate growth in the IFSC, but clearly it is significant in terms of the future development of the IFSC.
7. In October 1991, German government moves to recover tax from German companies in the IFSC gave rise to a wave of speculation with regard to the reactions of other governments and to lobbying of German and EC authorities by the Irish government.
8. To determine the actual impact of the IFSC on total employment it would be necessary to discount existing jobs transferred into the IFSC and any other negative impact of the IFSC on employment outside the IFSC.
9. The tax write-off might easily be higher. It is difficult to find hard evidence for

this important point. Anecdotal evidence (which emerged, for example, in press reports regarding the efforts of Dermot Desmond to unitize the South Block of the Custom House Docks) suggests that, ultimately, the tax revenue from IFSC operations might be in the region of IR£400–500 million. This is supported by evidence which suggests that the tax returned by the IFSC (at 10 per cent) for 1993–94 was in the region of IR£140 million. This would indicate a tax loss for that period, in relation to the corporation tax incentive, of up to roughly IR£450 million, depending on assumptions regarding the ratio of IFSC business subject to the 10 per cent incentive and assumptions regarding the tax that might be paid by companies in the absence of all incentives.

10. The initial capital cost is given as IR£400 million in the Annual Report of the CHDDA for 1989.
11. Although beyond the scope of this Chapter, it is interesting to note that the use of incentives may also promote occupiers into a more influential position in terms of property markets and development processes. Where urban development is fuelled by means of financial incentives targeted at (or routed through) occupiers, the traditional 'bottom-up' development process, which invests power with property interest, may be restructured into a more 'top-down' or demand-led development process, with some attendant changes in the pattern of relationships between occupiers and property or development interests. At another level, the use of incentives implies a partnership between the government and private development interests, and in that respect it might also be useful to examine how the situation for development interests is altered when the state has a greater involvement in the development process.
12. The American practice of Benjamin Thompson and Associates and the Irish practice, Burke Kennedy Doyle and Partners, are architects for the project.
13. Most obviously perhaps, where an investment interest, in pursuit of profit, fails to secure adequate rental income or a capital gain.

BIBLIOGRAPHY

Ahern, B. (1994) 'Government policy and the latest urban renewal legislation explained', paper presented at the conference City Living in the Year 2000 (May), Dublin.
Benson, F. L. (1988) 'The Custom House Docks', in J. Blackwell and F. Convery (eds) *Revitalising Dublin: What Works?*, Dublin: Resource and Environmental Policy Centre, University College Dublin.
—— (1991) 'Public/private sector partnerships: the Custom House Docks: a case study', paper presented at the conference Cities on Water, Second International Meeting, Venice.
—— (n.d.) *Planning for Development: the Custom House Docks experience*, Dublin: Custom House Docks Development Authority.
Cahill, G. (1988) 'Custom built', *Building Design* 873: 20–7.
Corrigan, K. (n.d.) *The Role of Fiscal Policy in Urban Renewal*, Dublin: Kieran Corrigan and Co.
Custom House Docks Development Authority (1987) *Annual Report and Accounts*, Dublin: CHDDA.
—— (1988) *Annual Report and Accounts*, Dublin: CHDDA.
—— (1989) *Annual Report and Accounts*, Dublin: CHDDA.
Department of the Environment (1986) *Urban Renewal Financial Incentives*, Dublin: Department of the Environment.
—— (n.d.) *Construction Industry Review '92, Outlook '93*, Dublin: The Stationery Office.
Ernst and Whinney (1987) *Ireland as a Haven for International Banking and Financial Services*, Dublin: Ernst and Whinney.
Government of Ireland (1993) *National Development Plan 1994–1999*, Dublin: The Stationery Office.
—— (1994a) *Programme for Competitiveness and Work*, Dublin: The Stationery Office.
—— (1994b) *Finance Bill 1994*, Dublin: The Stationery Office.
—— (1994c) *Finance Bill 1994: Explanatory Memorandum*, Dublin: The Stationery Office.
Hardwicke, McInerney and British Land (n.d.) *Custom House Docks Development Submission*, Dublin: CHDDA.
Harvey, D. (1989) *The Condition of Postmodernity*, Oxford: Basil Blackwell.
Kavanagh, M. (1991) 'The anatomy of a property company', unpublished MA thesis, University of Dublin, Trinity College.
Malone, P. (1990) *Office Development in Dublin 1960–1990*, Manchester: University of Manchester.
—— (1993) 'The difficulty of assessment: a case study of the Custom House

Docks, Dublin', in K. N. White, E. G. Bellinger, A. J. Soul, K. Hendry, M. R. Bristow (eds) *Urban Waterside Regeneration: Problems and Prospects*, London: Ellis Horwood.
McDonald, F. (1985) *The Destruction of Dublin*, Dublin: Gill and Macmillan.
McLaran, A. (1993) *Dublin: The Shaping of a Capital*, London: Belhaven/Pinter.
Sullivan, A. (1995) 'Special relationship brings U.S. funds business to Dublin', *International Herald Tribune*, 23–24 September: 17.

5

SYDNEY: THE ECONOMIC AND POLITICAL ROOTS OF DARLING HARBOUR

Maurice Daly and Patrick Malone

INTRODUCTION

The 1980s marked a turning point for the Australian economy. Australians gradually perceived that the economic structure that had sustained the country through most of the twentieth century would not continue to generate previous levels of prosperity. The public sector was converted to the principles of economic rationalism. The buzz-words of 'global best practice' and 'efficiency' heralded new economic policies that favoured market freedom over state involvement. For the government, the mechanisms of privatization and the exhaustion of 'hollow-logs' (unexploited government reserves) became the favoured means to achieve new economic goals. Private enterprise, fostered by alliances between politicians and entrepreneurs, was stimulated by fiscal and tax policies.

Darling Harbour emerged against the background of Australia's new economic and political order. It symbolized the new financial economy, Sydney's rise in the national and global financial markets, and a specific period in Australia's history. Located on a 54 ha site adjoining the western edge of Sydney's central business district, Darling Harbour was Australia's largest waterfront redevelopment (Fig. 5.1). As Farrelly (1989) points out, the area is bigger than St James Park in London or Copenhagen's Tivoli Gardens. Moreover, a number of other sites have since become part of what the public see as Darling Harbour, and a string of hotels are now physically linked to the original site.

Darling Harbour represents a specific phase in the redevelopment of Sydney's waterfront. It can be examined in terms of broad economic forces that reshaped the inner city, and factors that have had a direct bearing on Sydney's ports and dockland areas. The factors that rendered Darling Harbour redundant as a port and brought about the relocation of port facilities are common to other ports around the world; notably, changes in port facilities and shipping (Hoyle *et al.* 1988; Hall 1991). Similarly, the processes of redevelopment are comparable with those in other waterfront areas. In

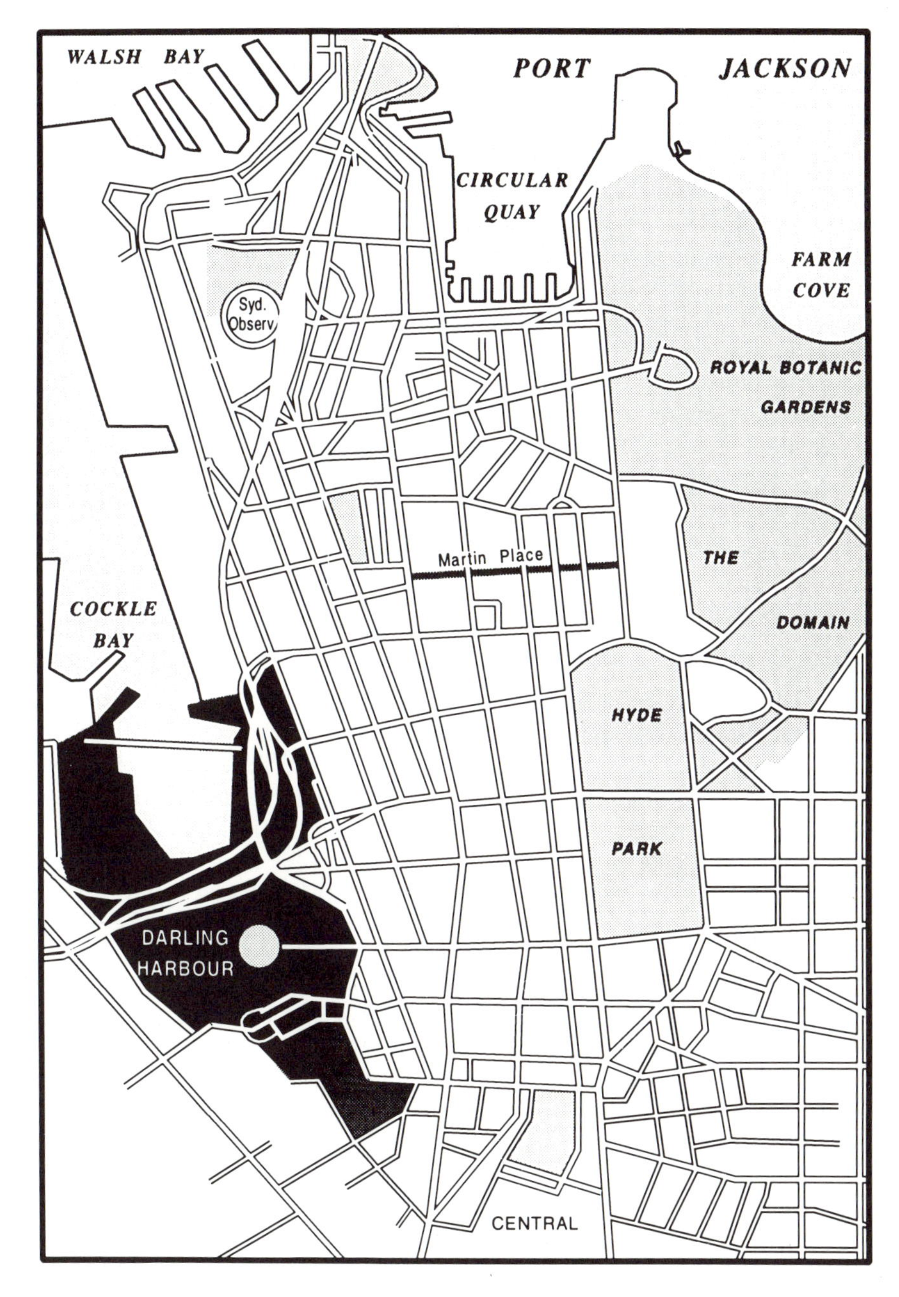

Figure 5.1 Sydney: CBD and Darling Harbour

common with other waterfront projects, Darling Harbour has provoked conflict over social and economic goals. There have been allegations of collusion between public and private interests. There was a public outcry regarding particular features of the scheme. Initial promises of public revenues and jobs gave way to the characteristic scramble for construction and development profits. The project provided an arena for the battles over property rights and the distribution of financial rewards. Economic returns, which were to be financed by the public purse, failed to materialize and public expenditure was higher than anticipated. Moreover, in common with other major projects around the world, Darling Harbour was clothed in marketing and media hype and the symbolic paraphernalia of postmodernist architecture. However, the economic and political forces behind the project were particular to Sydney and Australia in the 1980s.

Darling Harbour reflected key economic and political changes. It marked the rise of a monetarist economy rooted in financial and property capital and, in this respect, it is comparable with London Docklands. It has been described as an 'audacious piece of electioneering' (Farrelly 1989: 63), but it was also an icon of political, cultural and economic conditions. It symbolized structural changes in Australia's economy, Sydney's bid for global status, as well as the personal ambitions of specific politicians. Its role in the 1988 Bicentenary of the settlement of Australia by Europeans enhanced its symbolic value. But for some observers it also raised doubts about the future of Australia at a key point in its development.

The significance of Darling Harbour as a symbol lies partly in its location. As Morrison notes, Sydney Harbour is the 'generative force' in the city. It is 'central to Sydney's idea of itself' (Morrison 1991: 3). The nature of Sydney's central area and the character of its citizens make the waterfront an important physical element in the city and a major social forum. In the absence of a tradition for monumental or 'civic' planning, urban design and marketing values are vested in Sydney's harbour rather than in the city centre. Waterfront projects have a special significance, particularly when they increase public access to the water, reconnect the city to the harbour, or promote coherence in the city's spatial structure. Sydney can be compared with other cities where the juxtaposition of the core and the waterfront is immediate and dramatic. But Sydney's waterfront has a special significance as the city's 'living room' and a stage on which Sydney and Australia are promoted to the world. In effect, the rituals of the city and the state are played against the backdrop of the waterfront, Darling Harbour, Harbour Bridge and the Sydney Opera House (Andersons 1991: 11). However, as a national and civic monument, the status of Darling Harbour has been subject to change. Many of the political and economic forces which created it, and which it was intended to symbolize, have fallen from grace or passed into history. Many of the entrepreneurs associated with Australia's 'new economy' are now bankrupt, and some of the state's politicians are in gaol. Other politicians associated with the project have taken up lucrative positions in the private sector. Federal politicians who helped to launch the economic strategies of the 1980s now find that recession has destroyed the public's confidence in their policies.

This chapter traces the roots of Darling Harbour within the Australian economy and the economic and political situation in Sydney in the 1980s. It examines Darling Harbour on the basis of the circumstances surrounding its development and as a finished project.

DARLING HARBOUR: THE BACKGROUND

There was a marked increase in the size of the financial sector of the Australian economy in the 1970s and 1980s. Its growth was aided by a policy of increasing deregulation, particularly in the period 1984–86. The rise of the financial services industry was reflected in higher levels of lending and, for example, a wave of mergers and acquisitions in the second half of the 1980s. Attention shifted from manufacturing to financial capital while, socially, changes in the economic structure were mirrored in rising unemployment and part-time work, lower wage levels and greater disparities in income.

While Australia's policy-makers tried to meet the challenges of the global economic processes which emerged in the 1970s, they were confronted by two uncomfortable facts. First, that the changes occurring in Australia's internationalized economy defied the rules of post-war economic policies. The Keynesianism which dominated policy-making prior to the 1970s was founded on the assumption of economic growth. The international economic framework provided relative stability for world trade.[1] Fiscal measures were adopted to dampen the economy in periods of development and to stimulate growth in periods of recession. Production and consumption were stimulated by Keynesian tactics, and the government interfered to offset fluctuations in the economy by manipulating factors such as wages, access to housing, and energy prices.

Given the presumption of growth, urban planning in the Keynesian economy revolved around concepts of efficiency, social equity, land use and urban resources. As growth seemed inevitable, attention was directed to the size of cities (the optimum city debate) and location and patterns of urbanization (the growth pole debate). In the 1970s, however, these concerns faded. The assumptions regarding economic and urban growth weakened in the face of a new international economic order and fundamental changes in the role of agriculture and manufacturing. It became more difficult to predict the outcome of global processes or to control them within a given urban area. New limitations were placed on the ability of policy-makers, and particularly local policy-makers, to regulate the economic system and its spatial consequences.

Darling Harbour emerged in this phase of national economic and social change and shifting planning perspectives. In Sydney, the population and industrial base of inner-city areas had been declining for a number of years. Sydney's land-use pattern reflected the restructuring and spatial redistribution of industrial and retail functions (Law 1988). The new financial economy led to other dramatic changes, particularly within the core, where a rash of new office developments demonstrated the flow of financial and property capital into Sydney (Daly 1982). Sydney's waterfront also expressed fundamental changes in Australia's economy. Whereas Darling Harbour had served as a major exit point for primary exports (especially wool), its decline reflected a reduction in the flow of products to the port.

Two interrelated factors underpinned the redevelopment and expansion of Sydney's commercial core and spurred the initial invasion of the old port area: the general expansion of the financial sector and its concentration in Sydney, and the associated surge of development capital into the city. The growth of Australia's financial sector was initially triggered by the 'mineral boom' of the late 1960s, which fuelled investment and activity in the stock market. Foreign banks (although allowed to operate only as merchant banks) flocked into Australia. Roughly a hundred new banks were established between 1968 and 1972. Australia's access to burgeoning European markets further accelerated the growth of the banking system and, with the fall of the Bretton Woods Agreement (which had segmented international capital markets since 1944), Australia was poised to tap the flow of international capital. The 1970s and 1980s brought a chain of events that led to Sydney's emergence as a world city and the redevelopment of its central business district.[2] As twenty-four-hour trading became a reality and cities vied to assume a significant role in the world market-place, the New South Wales government was quick to appreciate Sydney's potential. Sydney could not hope to match the foremost world financial centres. However, it could attempt to establish a strong niche in the system of world cities. Financial deregulation had provided greater access to domestic and international markets. The new spirit of innovation led to the opening in Sydney of the first international futures market in South East Asia. With the influx of foreign banks, the city came to represent 'new capital' – as opposed to the 'old capital' of Melbourne. It exploited its position as the headquarters of the Reserve Bank. It marketed its harbour views and cosmopolitan lifestyle as the first invasion of international banks, and subsequent growth in the financial sector, were followed by a second invasion in 1979–83.

Globalization and the structural changes which took place in the economy were not clearly understood. Whereas earlier governments were seen to have misread the economy, in the 1980s Australia chose a new orthodoxy akin to that of Reagan and Thatcher. In 1982, the incoming Labour federal government placed its faith in new policies and sought to distance itself from previous Labour economic programmes.[3] It promoted partnerships with business, removed constraints and regulations and embraced managerialism in the public sector. Labour's new criteria were set in terms of the efficiency of the market-place.[4] Financial services and property became the growth industries. In the period 1980–91, employment in the financial sector more than doubled (to 344,000) so that the number of firms in the financial sector outstripped construction by 68 per cent and manufacturing by 147 per cent (Census 1991). Growth in the financial sector was paralleled by the flow of capital into urban development and property markets. The expansion of both sectors was boosted by the decline in industrial investment opportunities. Moreover, the financial and property sectors were mutually supportive. The enlargement of the financial sector underpinned the demand for new office space and, in turn, opportunities for investment in property were made available to a burgeoning financial sector.

Sydney's property boom was also augmented by the Australian reverence for property and its deep faith in property as an abiding source of wealth.[5] But the new economy brought new problems. Financial deregulation seriously damaged Australia's current account position in the second half of the 1980s and early 1990s. Following the push towards deregulation (1984–86) the banks and many of their clients became seriously overextended. When the stock market crashed in 1987, property was boosted as a major target for investment funds. This generated an investment-led boom, for which the banking system paid the price in the early 1990s. When interest rates were increased in 1989 there was a general shake-out in the economy and companies fell both within and outside the financial sector.

Eventually, the recession of the early 1990s brought a halt to the heady growth of the financial sector. Rising investment pressure and emerging difficulties in the Australian economy led to a desperate search for new industries, preferably ones that would generate substantial export earnings. Thus, tourism emerged as a major new element of the economy – a fact that would be reflected in the redevelopment of Darling Harbour.

The rise of tourism must be understood in the light of deregulation. During the 1980s, Australia's current account deficit rose from $A5.5 to $A22.3 billion. This was due not to Australia's trading performance but to the outward flow of capital, which roughly doubled over the decade until, in 1991, it represented 75 per cent of the trade deficit.[6] In the ten years to 1991 Australia's gross debt increased from 11 to 44 per cent of GDP. There was an explosion in private debt in the mid-1980s: with the lifting of capital and exchange controls, money flowed into and out of Australia without restraint. Controls over interest rates and reserves were also removed and full banking licences were issued to a number of foreign banks. These measures were taken on the assumption that the market-place would provide a competitive, disciplined and efficient banking system. However, over the decade, private foreign debt rose from $A8 to $A128 billion. Private domestic debt, three-quarters of which involved financial institutions, rose to $A458 billion (EPAC 1991). Borrowings were commonly fed into company take-overs, mergers and property. Over the decade to 1991, the share of foreign debt represented by these outlets increased from one-sixth to roughly half. As debt was predominantly short term, interest payments mounted until, in 1990 for example, interest payments represented roughly 64 per cent of gross foreign debt. Meanwhile, the surge in debt boosted inflation to levels between twice and ten times those of Australia's trading partners.

Ultimately it became apparent that Australia could only withstand its economic problems by attracting foreign investment and boosting foreign earnings through new and existing industries.[7] The ensuing flow of foreign investment was highly concentrated, both in terms of the countries that invested in Australia and the economic sectors which attracted investment. While Japan was the major single investor from 1986, Japanese investment was concentrated in real estate and tourism. Significantly, roughly two-thirds of Japanese real-estate investment was in Sydney and New South Wales.[8]

While Japanese direct investment helped Australia's economy through the critical years of the late 1980s, tourism, and especially Japanese tourists, provided a general boost to foreign earnings.[9] Following the first National Tourism Outlook Conference (Canberra, 1981), expenditure on tourism increased by an average of 12.3 per cent annually throughout the 1980s. Foreign earnings from tourism roughly quadrupled between the mid-1980s and 1992 until, eventually, tourism became the largest economic sector in terms of foreign earnings. The New South Wales Tourism Commission was established in the same year as the Darling Harbour Authority (1984).

DARLING HARBOUR

In the late 1980s, Darling Harbour echoed shifts in the Australian economy as tourism emerged to redress the loss of industrial employment and the balance of payments problems engendered by Australia's new economic structure. The project was also a political show-piece, used to demonstrate the efficiency and managerial skills of the New South Wales government and the status of its premier. In addition, it was an exercise in property development. As a model of the new monetarism, it was to return a profit to the public purse and to pave the way for private enterprise without placing a burden on the taxpayer.

For a while, Darling Harbour became a national icon, as Australia searched for new symbols against a background of uncertainty – uncertainty generated partly by the failure of its economic theories and a growing awareness of its complacency regarding the tenuous conditions that had, until recently, secured its wealth. Unfortunately, however, the pressures on Darling Harbour to symbolize a new age meant that the project was unlikely to be responsive to wider social criteria. The history of the government's involvement with comparable projects also suggested that social idealism would have no place in Darling Harbour. Previous waterfront projects generally favoured commercial development and private interests. They promoted social conflict because successive New South Wales governments, of whatever political persuasion, preferred development to conservation and commercial values to social needs. This was hardly surprising, given the wider social attitudes to property and the part played by the government in fostering those attitudes.[10] In the 1960s, for example, the government set up a special authority to redevelop The Rocks, a historic area abutting the north-west corner of the commercial core. This authority drew planning powers away from the city and, in the euphoria surrounding the late 1960s property boom, it hatched a plan for a large-scale commercial development with a core of high-rise office buildings. The proposals for The Rocks, and another redevelopment project for Woolloomooloo (an old port and residential area north-east of the core) failed because of strong opposition from an unlikely coalition of trade unions, working-class residents, and middle-class conservationists. Workers placed 'green bans' on construction sites and the developments prompted public protests that were sometimes violent.

In the affluent years of the early 1970s, it appeared as if a new age of preservation was dawning in Sydney. The federal government intervened to provide working-class housing in Woolloomooloo. The original plans for The Rocks were substantially abandoned – although in the period after 1974 many commercial development companies and development proposals were scuttled by the collapse of the property boom.

The anti-development successes of the early 1970s resulted in some international acclaim for Sydney, but this was short-lived. In June 1984, a decade after the battles of The Rocks and Woolloomooloo, Neville Wran, the Labour premier of New South Wales (1977–87), introduced the Darling Harbour Authority Act (1984). An authority was created to circumvent established planning procedures and to stifle opposition. It gave planning and development powers to a board charged with the rapid redevelopment of Darling Harbour. The creation of the special development authority also put paid to previous plans, drawn up by the state's Department of Environment and Planning in 1982, for a mixed residential and recreational development on the site of Darling Harbour.[11]

In 1984 there was remarkably little opposition to the government's plans for Darling Harbour. Arguably, it was difficult to formulate specific objections. The rusting railway sheds had little appeal. There was no resident population, and the need to revitalize traditional working-class industrial areas failed to excite a broad public reaction. Moreover, the public may have been impressed by the project's symbolic value to Sydney as a world city and an international financial centre. The use of imagery that played on the project's civic purposes may have excited less resistance than the overt commercial objectives of earlier waterfront projects. Moreover the project was launched alongside a number of other high-profile developments which emerged in Sydney as part of the lead-up to the 1988 Bicentenary. Under Premier Neville Wran, these included new sports and cultural facilities, public gardens and major additions to the Art Gallery of New South Wales and the Australian Museum. Historic buildings were restored. Sydney's partly defunct passenger port (Circular Quay) was redeveloped.[12] However, Darling Harbour was the centrepiece of Wran's preparations for the bicentennial celebrations.

When he first turned to the site in the late 1970s, Wran's ambitions were frustrated by the reluctance of the Conservative federal government to share high-profile projects with a state Labour government. In 1980, however, Wran channelled ambitions for 'Expo 88' towards the site with the intention that this would form part of the 1988 Bicentenary celebrations. In 1983, with the backing of the newly elected federal Labour government, Wran turned again to Darling Harbour. His original concepts for the site were restructured to reflect Sydney's bid to become a global financial centre.[13] The emphasis was on the provision of international conference and exhibition facilities but, as described at the time of the official launch on 1 May 1984, the scheme also included a high-technology 'discovery village' for family entertainment, an aquarium, 'festival' retail markets, waterfront promenades, a maritime museum, parks, a Chinese garden and a hotel and casino complex. At that time the casino was an important economic element in the project, but tourism, rather than gambling, ultimately provided the backbone of the project (Fig. 5.2).

At the time that the project was launched in 1983–84, it was not clear that tourism would eventually rival the financial sector in terms of foreign earnings, or that Sydney would manage to gain more than one-third of Australia's foreign earnings from tourism. Arguably, the concepts for Darling Harbour could have been more finely tuned if the growth in tourism had been predicted more clearly. Nevertheless, tourism added an important dimension to the Darling Harbour project. It provided its economic core and enhanced its value to the city.

One of the primary goals of the project (as stated in the 1983 preparatory report for the Premier's Department) was to maximize financial returns to the government (Premier's Department 1983: 2). It was initially estimated to require $A200 million in public expenditure. By the time that detailed plans were unveiled in December 1984, the project had become a joint public/private venture with an estimated cost of $A1 billion. Wran's claims for the project had grown in line with the estimate of costs. He described it as a bicentennial gift to the nation and as the greatest urban redevelopment ever undertaken in Australia. To ensure that a gift of such

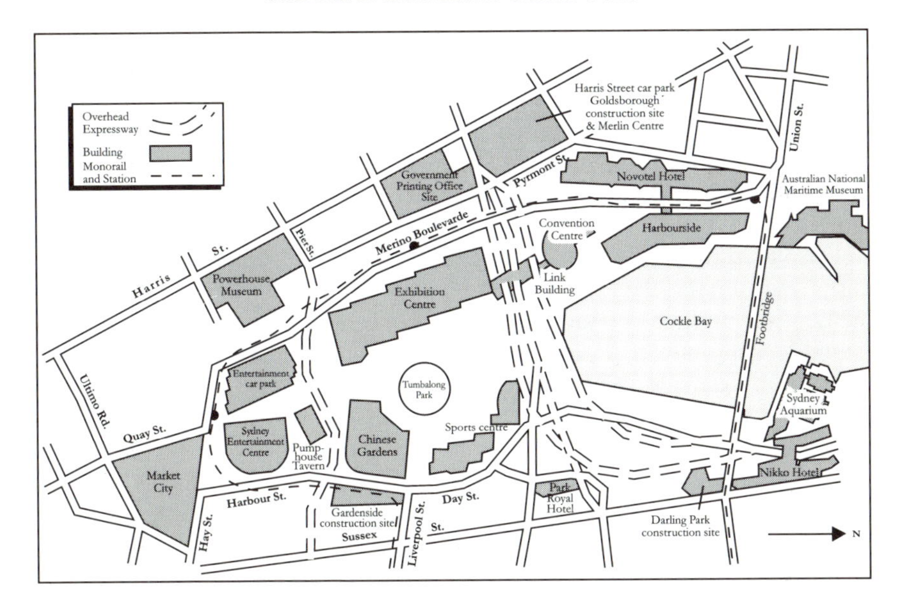

Figure 5.2 Sydney: Darling Harbour site, 1992

magnitude was properly presented, the Darling Harbour Authority was given powers that superseded those of the City Council, the National Trust and the State Planning Department. Essentially, the project's managers were accountable to the Premier and his designated minister, Laurie Brereton.[14] Given that Darling Harbour was to boost Sydney's position as a world city, it was argued that it should resemble comparable foreign developments. Its creators turned to the American pattern for waterfront developments. Baltimore provided a key example for a project that was to rank with the Opera House and Harbour Bridge as one of the three most important projects of twentieth-century Sydney (*Sydney Morning Herald*, 15 December 1984). The project was also to create 30,000 jobs while under construction and a further 10,000 jobs on completion. The deadline for completion was Australia Day in the bicentennial year (26 January 1988).

Darling Harbour was pushed forward at a dazzling pace. The four companies chosen to bid for the management of the project were given just one week to prepare their tenders. The pressure was such that the first chairman of the Authority resigned after ten days in office. Nevertheless, while chairmen might come and go, Wran believed that the Authority's chief executive (previously the manager of the Cooper Basin mining project) would bulldoze the development through. For its part, the government feared that the trade unions might use the Premier's desire to build a monument to himself to hold the government to ransom. This fear was not unfounded for, during an earlier Wran undertaking (the Sydney Entertainment Centre), union pressures for pay increases had been based on the claim that the smell of food from the nearby China Town promoted hunger among workers. To offset potential union problems at Darling Harbour, Jack Ferguson, a former Deputy Premier and previously an official of the Building Workers Industrial Union, was appointed deputy chair of the Authority.

Despite the government's precautions, union troubles surfaced at an early stage in the development. As Laurie Brereton (Minister for Public Works) was announcing an agreement between employers and unions that would 'ensure industrial peace and harmony and guarantee completion for 1988', the Waterside Workers Federation threatened to strike over the 'land grab' that allegedly threatened their jobs (*Sydney Morning Herald*, 9 August 1985).

The unions were not the only potential source of resistance to the project. In 1985, the Institute of Australian Planners criticized the transport planning associated with the site. In addition, Sydney's City Council planners reported that 'at the local level the proposal raises major environmental and planning impacts in terms of traffic, transportation, land use, design, accessibility and social issues' (*The Australian Financial Review*, 7 February 1985). The criticisms raised by the planners were ignored and the entire City Council, which had vigorously opposed the development, was eventually sacked by the state government.

There were also allegations that the government had handed lucrative building contracts to favoured companies. The rapid pace of the development process added substance to speculation regarding collusion between public and private interests. Suspicions were also aroused by some unwitting remarks made by the second chair of the Authority, when he reported that 'for years before the Authority was even set up officials from the Premier's department were having regular discussions with major companies' (*Sydney Morning Herald*, 4 March 1985).

There was also widespread criticism of the lack of any residential element in the proposed scheme. Scepticism was fuelled by the marked contrast between the project and the run-down working-class areas bordering its site. Laurie Brereton countered this line of argument by claiming that an 'immensely attractive prestige development' in Darling Harbour would boost adjoining residential areas and 'attract renovation and renewal' (*Sydney Morning Herald*, 4 March 1985).

The mounting problems that confronted Darling Harbour spurred the government's ambition to speed up the development process. The first stage of the project was to consist of the exhibition centre, the convention centre, Chinese garden, maritime museum, and 'festival markets' – developments with a total value of $A450 million. In 1985 Brereton put pressure on the developers by declaring that no part of the project would be allowed to start unless it could be finished in time for the Bicentenary (*Sydney Morning Herald*, 9 August 1985).

While the twin forces of marketing and a pressurized development process were employed to 'jump start' Darling Harbour, the spectacle, glitter and extravagance of the project were meant to dazzle the electorate and to convince voters that the Labour government was destined to achieve its goals. Initially, the project was generally well received. This was due to Wran's popularity, the gloss surrounding the project and the fact that it had a symbolic value in terms of Sydney's status. In this respect, it is interesting to compare the relatively warm initial reception awarded by the public to Darling Harbour with the negative reactions to earlier development proposals for The Rocks. Whereas the proposals for The Rocks were blatantly commercial, the fact that the economics of Darling Harbour hinged initially on the potential profits of a casino was partly camouflaged by the symbolic value of the project. However, as an editorial in the *Sydney Morning Herald* (9 August 1985) observed: 'building unions, architects, public money and political deadlines are a potentially lethal combination [and] Darling Harbour could easily become an electoral liability to the Government'.

Massed public opinion began to turn against Darling Harbour in 1986. Public dissatisfaction was focused on the potential cost of the project to the public purse and on proposals for a monorail. The monorail was described by the Transport Department as an 'attraction' rather than part of the public transport system. Privately funded, it was legitimized by a special act of parliament that made it exempt from normal planning controls. Raised above street level, the monorail was routed along the perimeter of Darling Harbour and through part of the commercial core. The negative effects on streets and valued older buildings gave rise to two protest marches. A letter to the *Sydney Morning Herald* typified the reaction of the general public:

> The monorail will only serve to further deface the city of Sydney. However, my major concern is not with the aesthetic damage which will be done to the city but with the high-handed way in which its residents have been treated. It is a farce for anyone to maintain that this is a democracy when Australian citizens are dragooned into accepting such totally unnecessary monstrosities in their midst. The government deserves to lose office over this issue.
>
> (*Sydney Morning Herald*, 31 July 1986)

Whereas the public's reaction to Darling Harbour hinged to a great extent on the monorail, it could be said to have triggered a general reaction to the 'the ill-conceived insertion of gigantic new developments into the city's fabric' (Morrison 1991: 3). The monorail symbolized the government's arrogance and the tendency to put political advantage and private economic ambitions before essential development and the public interest. The fact that the monorail was to be built by Sir Peter Abeles, a friend of the federal Labour Prime Minister and head of the largest transport group in the country, gave rise to allegations of 'cronyism'. Thus, for a number of reasons the monorail was a political 'hot potato' which, following Neville Wran's resignation from politics in 1986, fell into the lap of Premier Unsworth.

Another factor that evoked a strong public reaction was the selection of the Hooker–Harrah consortium to build a hotel and casino in Darling Harbour. This part of the project, which was to cost $A750 million, was to be the principal source of revenue through which the government would recover its $A450 million investment in Darling Harbour. In an attempt to allay criticism of this level of public expenditure, the State Treasury produced a cost–benefit analysis which predicted that the government would reap $A100 million a year in gambling revenue from the casino. On this basis, it was claimed that public expenditure would be recovered within six years. In 1987, however, following allegations that the Harrah group had underworld connections in the United States, the government was forced to revoke the approval given to the consortium to build the casino/hotel. This provided the basis for another scandal. With the abandonment of the casino, and in the absence of a comparable source of revenue, it became clear that the public purse would be left to carry the state's $A450 million investment in Darling Harbour. This prompted the *Sydney Morning Herald* (31 July 1986) to remind the government that 'countries with large, long-term balance-of-payments problems just cannot afford to squander fortunes on public parks and monuments'.

In October 1987, the government was forced to accept that Darling Harbour would not be finished for its formal opening in 1988. The Premier blamed industrial disputes while, in the same month, the Minister for Public Works sacked 1,400 construction workers on the Darling Harbour site, supposedly because of a demand by the Electrical Trade Union for a 4 per cent pay rise. These difficulties, and public reaction against Darling Harbour, contributed to Labour's defeat in the 1987 election. The government was accused of 'profligate government expenditure' (Andersons 1991: 11). The ambitions which Wran had invested in Darling Harbour, and the pressures imposed by Brereton to force the pace of development, came to nothing in political terms. The various economic and political returns envisaged in 1983–84 proved elusive. When Queen Elizabeth opened Darling Harbour in May 1988 the half-finished project was in the hands of a new Conservative government. The remainder of the site took another four years to complete.

After a change of government and the general euphoria generated by the Bicentenary in 1988, the public's reaction to Darling Harbour softened. The launch of the half-completed project was greeted with some enthusiasm. The project was seen as an appropriate addition to the city centre at a point when Sydney was going through another property boom (triggered by expansion of the money supply and efforts to dampen the effects of the 1987 stock-market crash). The Bicentenary, and an innate reverence for big urban developments, helped to establish Darling Harbour as a potential tourist landmark.

The public's acceptance of Darling Harbour may also have been encouraged by the fact that the change of government in 1987 paved the way for a new emphasis on planning and greater co-operation between the state and city authorities. The Central Sydney Strategic Plan, launched in 1988, embodied an attempt to provide an overall plan for Sydney's beleaguered urban structure. It built on the city's landscape and topography to integrate disparate elements within a network of physical and visual linkages. The plan also looked to the expansion of the commercial core to reinforce linkages between the core and the waterfront. It identified the Ultimo Pyrmont area to the west of Darling Harbour as a 'special development area' for residential, tourism and commercial use. Thus, it reinforced the importance of the waterfront and paved the way for further invasion of the waterfront area.

ASSESSING DARLING HARBOUR

It is clear that Darling Harbour was shaped by a number of factors. It was moulded by national and local politics and economic events. It reflected a particular situation in Sydney in the 1980s. However, there is no simple yardstick for its success or failure. The project symbolizes different things to different people. With its borrowed postmodernist architecture, Darling Harbour may be seen as an icon of Australia's spend-thrift economic policies and political brinkmanship – a grand gesture which cost $A500 million in public debt. For some, it may be a monument to the spurious economic policies of the 1980s, which eventually submerged domestic and foreign banks in debt. It may symbolize reckless spending on property, failed entrepreneurs and foolhardy investment, or the vacant and half-cleared sites, abandoned developments and empty office space of Sydney's commercial core in the early 1990s.

There are, however, more generous views of Darling Harbour (Fig. 5.3). It is now a functioning element in Sydney's economic and cultural landscape and has won guarded approval from abroad (see for example Quarry 1987). For some Australians, Wran's intentions in instigating the project went beyond narrow commercial and political ambitions. They argue that he had a broader vision. Following his retirement in 1986, he was praised in an editorial in *The Australian* newspaper for his leadership in urban regeneration. He is said to have exploited Sydney's potential, its rich cultural facilities and memorable image, and to have promoted tourism and the city's economy. In this context, Darling Harbour is seen as an expression of his

Figure 5.3 Sydney: Darling Harbour Exhibition Centre
Source: Architectural Review 1106, April 1989

success (Andersons 1986). Thus, Darling Harbour demonstrates that projects are open to interpretation. For many, it makes a positive contribution to the life of the city. For the cynical, or perhaps the more discerning, Darling Harbour is an 'escapist fun zone', so far removed from the ordinary functions of the city that even Sydney's citizens are turned into tourists (Farrelly 1989: 65).

Clearly, Darling Harbour is more likely to be seen as a positive contribution to the city and society if assessed as a completed development. It may win approval as an urban landmark or as a tourist attraction. Positive reactions are likely to be based on its physical and functional characteristics, while those who know its history may see the forces behind its development in a negative light. Critics may see evidence of narrow or private economic and political ambitions, or even of the Australian predilection for reckless property development. It is significant that Darling Harbour was to return a profit to the government through the operations of a casino, and that this failed to materialize because of the peculiar *naïveté* of Australian politics and the fear of fundamentalist voters. A bluff and commercial approach to urban development was undermined by the discovery that gambling attracted the heavyweights of international crime. It is also significant that, without its cash-cow, Darling Harbour passed half a billion dollars of public debt on to future generations.

Darling Harbour was a notable exercise in state-sponsored aggression which set aside institutional planning restraints and marginalized its critics. Normal procedures for large-scale projects, such as cost–benefit analyses, were disregarded until public disquiet forced the Treasury to issue a face-saving and patently incorrect estimate of costs and returns. But some interests associated with the project also lost out, and the alliance of government and commercial interests had unfortunate consequences for both parties. The project was marred by squabbles over land ownership and compensation. The first major developer ended up bankrupt. The small retailers who obtained expensive leases at the height of the property boom were left to struggle through the ensuing recession. The hotels that were introduced to fill the void created by the abandonment of the casino have struggled, during the recession in 1991–92, with occupancy rates of 40 per cent or less.[15]

It is also significant that the impact of the project on surrounding areas was ignored. For the working-class areas to the west there were hopeful expressions of some trickle-down effects. In the central city area to the east, the monorail represented an unwanted intrusion. But the monorail was a significant 'hinge' in the project's history. It epitomized the government's bullying tactics. Together with the project's financial 'black holes', it triggered an adverse public reaction which hastened the downfall of the Labour government.

Given these issues, Darling Harbour might be considered a failure in terms of some of its economic, political and social consequences. Yet, by another accounting procedure, it is successful. The controversies that surrounded its development will fade from collective memory as the completed project is increasingly perceived as a physical and a functioning element of Sydney and a useful addition to the tourist's checklist of attractions in a period when tourism is the largest contributor to Australia's foreign-exchange earnings. It now absorbs well over a million visitors a month. Its value to Sydney's central city area is less obvious. Its 'urbanity' is in doubt. Some critics argue that its ethos is anti-urban or suburban and that 'it is simply less a part of the city than a refuge from it' (Farrelly 1989: 63). On the other hand, it has increased the public's access to the waterfront, which is a major factor in the life of Sydney.

Whatever the balance of positive and negative reactions to Darling Harbour, one of the lessons to be drawn from the project is that a completed development may win public support while the motives of its creators, its political genealogy and the economics of its development may draw broadly negative reactions. In short, successful projects may spring from dubious social, political or economic development processes. As a completed development, Darling Harbour has been interpreted as an expression of Sydney's commitment to 'an open, expansive and egalitarian environment' (Morrison 1991: 3). Here, the images of an insensitive development process driven by self-serving politicians have given way to images of an open and democratic society.

It is important to note, however, that the reinterpretation of a completed project does not invalidate negative interpretations of the development processes behind that project. It simply means that it is important to distinguish between the examination of development processes and of completed developments. Moreover, the key determinants of urban development – the 'lower layer' of political and economic ambitions – may remain fairly constant. Some 4 km north of Darling Harbour, at Pyrmont, Sydney's development industry has focused on the large abandoned waterfront site of the former Colonial Sugar Refinery complex. While a casino is being built on land once occupied by an old power station, this site is the focus of old controversies between commercial interests and those pressing for housing and parks. Meanwhile, to the west of Sydney, preparations for the Olympic Games have meant that urban development can be hung on a new social peg. As the struggle between social and commercial values in Pyrmont and the political point-scoring that surrounds the Sydney Olympics development suggest, urban development in Australia will continue to adopt a particular political and economic character. But in this respect it should also be noted that, whereas institutional restraints were marginalized in Darling Harbour, the ultimate curbs of democratic government remained in place.

NOTES

1. Arrangements such as GATT and the Bretton Woods financial agreement were engineered to foster a liberal trading regime. The compliance of nations with these arrangements depended on the hegemony of the US in world economic

affairs. As the US position weakened, so too did the interest and ability of various nations in maintaining the arrangements. Indeed, some of the most successful growth nations of the period, like Japan, derived part of their advantages exploiting the opportunities that open access gave to the wealthiest markets (especially the US) while protecting their own markets, and manipulating exchange rates.

2. The Bretton Woods system was not able to survive because of internal flaws, most notably the centrality of the US dollar and its tie to gold. As other economies began to rival that of the US, the value of the US dollar was challenged. The challenges to the US economy signified a great expansion of world trade, and gold production could not be increased to keep up with it. A world of fixed financial relationships inevitably became a world of mobile relationships.
3. This was evidenced in 1973, when the federal Labour government decided to cut tariffs by 25 per cent and to open up the manufacturing base to international competition. Although this policy was partially reversed by Conservative governments in 1975–82, it accelerated the decline of the manufacturing base of the economy.
4. The New South Wales Labour government, which had survived the anti-Labour swings of the 1970s, provided lessons for the federal government in this regard.
5. There are historic factors behind Australia's devotion to property. Australia's entire 'European history' falls within the span of industrial capitalism. That history is one of cyclic patterns of growth: booms collapse into recessions as cycles of innovation, product development, capital accumulation and rates of profit interact over time. Uneven growth is amplified by the country's dependence on commodity markets and foreign trade. Historically, these markets grew when international trade (particularly with Britain) was buoyant. However, when external economies faltered, Australia's relative wealth induced counter-cyclic flows of capital and people (Thomas 1972). Where these counter-flows occurred *after* booms in agricultural or mineral trading, they tended to promote urban development and investment in property.
6. Although as a percentage of the trade deficit, trading fell during the decade from 36 to 13 per cent, Australia did not trade badly over this period. For half of these years the merchandise trade balance was in surplus. The record deficit of 1988–89 was only $A0.7 billion greater than the deficit in 1980–81.
7. The inflow of funds totalled $A37 billion from 1980 to 1984; $A91 billion in the four years from 1984 to 1988, and then $A31 billion in the three years up to 1991.
8. A further 27 per cent was directed into Queensland, where tourism is a major element of the economy.
9. Japan played a major role in both foreign investment and foreign earnings from tourism. In 1992, for example, 620,000 Japanese made up 24 per cent of the total number of visitors to Australia. Sydney benefited greatly in that, whereas more Japanese tourists went to Queensland (46 per cent) than to Sydney (32 per cent), Sydney drew the largest number of visitors from each of the other major source areas.
10. In the 1880s, opportunities for suburbanization afforded by the railways in Sydney resulted in the greatest land boom of the century: one that virtually broke the banking system and heralded a major depression. Thirty years later further expansion of rail and tram systems caused another boom in land values, which was halted by the Great Depression. The car-dominated expansions of the 1950s and 1960s created a generally steep and constant upward movement in

property prices. Thus, it is part of the folklore of Sydney that property represented a sound and profitable medium of investment.

11. This plan proposed high-density, medium-rise residential development around a central open space. In 1982, Alderman Jack Mundey, a leader of the Green Bans movement, also proposed a low-cost housing scheme for the site.
12. Circular Quay, situated roughly between the Opera House and the Harbour area, was redeveloped following an international conference in 1983 (entitled 'The City in Conflict') and a design competition organized by the Royal Australian Institute of Architects. The New South Wales Public Works Department proposed a project to upgrade the Circular Quay which was initiated in the mid-1980s under Premier Wran. The redevelopment provided a promenade and other facilities along the waterfront. Essentially a landscape project, it greatly increased access to the waterfront and relieved the barrier effect of the 1950s overhead rail and roadway, which lie between the Quay and the core. Unlike Darling Harbour, the Circular Quay project was essentially uncontroversial.
13. The new federal Labour government, which was dedicated to the internationalization of the Australian economy, planned sweeping economic reforms in the financial sector and taxation. In the event, financial deregulation was its most far-reaching achievement.
14. The initial development strategy for Darling Harbour was prepared by the New South Wales Public Works Department. The private firm of MSJ was appointed in 1985 as project design director. MSJ was responsible for the planning, architectural and landscape aspects of the project and project management. Individual buildings were designed by other practices. A Quality Review Committee, comprising practitioners, academics and others, was appointed as a watch-dog.
15. Occupancy rates of 63 to 64 per cent were common in 1995.

BIBLIOGRAPHY

Andersons, A. (1986) 'Architecture in the Wran Era', *Architecture Australia*, November: 3–8.

—— (1991) 'Waterfront redevelopments: Darling Harbour and Circular Quay', *Urban Design Quarterly* 39: 11–15.

Church, A. (1988) 'Demand-led planning, the inner-city crisis and the labour market: London Docklands evaluated', in B. S. Hoyle, D. A. Pinder and M. S. Husain (eds) *Revitalising the Waterfront*, London: Belhaven: pp. 199–221.

Clark, M. (1988) 'The need for a more critical approach to dockland renewal', in B. S. Hoyle, D. A. Pinder and M. S. Husain (eds) *Revitalising the Waterfront*, London: Belhaven: pp. 222–31.

Daly, M. T. (1982) *Sydney Boom, Sydney Bust*, Sydney: George Allen and Unwin.

Economic Planning and Advisory Council (EPAC) (1991) *The Surge in Australia's Private Debt: Causes, Consequences, Outlook, Background Paper No. 14*, Canberra: Australian Government Printer.

Farrelly, E. M. (1989) 'Out of the swing out of the sea, Darling', *The Architectural Review* 1106 (April): 63–9.

Hall, P. (1991) *Waterfronts: A New Urban Frontier*, Berkeley: University of California at Berkeley.

Harvey, D. (1989) 'Downtowns', *Marxism Today* (January): 21.

Hoyle, B. S., Pinder, D. A. and Husain, M. S. (eds) (1988) *Revitalising the Waterfront*, London: Belhaven.

Huxley, M. (1991) 'Making cities fun: Darling Harbour and the immobilisation of spectacle', in P. Carrol, K. Donohue, M. McGovern and J. McMillen (eds) *Tourism in Australia*, Sydney: Harcourt Brace.

Law, C. M. (1988) *The Uncertain Future of the Urban Core*, London: Routledge.

Morrison, F. (1991) 'Sydney! Sydney!', *Urban Design Quarterly* 39: 3–5.

O'Connor, K. (1990) *State of Australia*, Melbourne: Monash University National Centre for Australian Studies.

Premier's Department (1983) *Darling Harbour Redevelopment Financial and Organizational Study*, Sydney: Ferris Norton and Associates.

Quarry, Neville (1987) 'Darling development', in *The Architectural Review* 1080 (February): 70–4.

Thomas, B. (1972) *Migration and Urban Development*, London: Methuen.

Tweedale, I. (1988) 'Waterfront redevelopment, economic restructuring and social impact', in B. S. Hoyle, D. A. Pinder and M. S. Husain (eds) *Revitalising the Waterfront*, London: Belhaven: pp. 185–98.

6

HONG KONG: A POLITICAL ECONOMY OF WATERFRONT DEVELOPMENT

Roger Bristow

One much debated aspect of waterfronts is whether the resurgence of interest in their development potential during the 1980s was a unique occurrence, or merely the coming together of normal property development processes in particular places at a particular time. There is little disagreement on the theory that technological change in the port industry in the 1970s produced opportunities for development (Pinder *et al.* 1988: 248–52), but evidence suggests that it presented no more than special locational advantages and did not alter the general dynamics of the property development industry (Breen and Rigby 1994). The likely scenario is that general development inputs or constraints were modified. New visions were perceived, new opportunities were introduced and old ones enhanced, while existing urban economic relationships, development opportunities, or methodologies for implementation remained fundamentally unchanged. This is demonstrated in Hong Kong, where there has been an intimate connection between land and water since the foundation of the city in the 1840s (Pun 1991: 33).

When looking at waterfront development in Hong Kong, it is important to remind ourselves of the fundamental rules of property development in capitalist societies, which underpin the processes of such development. It is widely accepted that private property rights lie at the heart of the capitalist paradigm. Much of the philosophical argument, and many of the opposing theories of political economy (Kymlicka 1990), concern varied and contrasting views of the relationships between competing private objectives, those of the state, and the theory and reality of what might constitute community interest. The outcome of this debate may legitimize land-use planning and development control and the systems of relations that involve the state, community and private interests in the property development process. This has led other commentators to assert that the key to understanding the development process and its interaction with planning is 'the identification of interests in land, its use and development, and the processes through which interests are mediated within the institutional arrangements of the planning system' (Healey *et al.* 1988: xii). Thus, it is contended here that the choice of the waterfront in Hong Kong as a location for an airport, while it constitutes a special use, does not change the theoretical fundamentals of the development process. Final plans or development processes are therefore, as Kirk has put it:

> undoubtedly the product of a process of horse-trading between [planning] authority and private developers and/or owners. This process may not be easy to trace – partly because so much of it goes on behind closed doors, though there may be some leaks – and not least because its likely outcome will probably have been partly anticipated . . . from the outset.
>
> (Kirk 1980: 51)

When planning authority, developer and landowner are one and the same, as in Hong Kong, the focus shifts to the likely secretiveness of the development 'process' and the amount of influence that any outsiders may have on such a process.

It should not be any surprise, therefore, that much of the discussion here about waterfront development in Hong Kong will focus on such interrelationships, and the various compromises and bargains arrived at by competing interests. But before examining these, we have to remind ourselves of a further basis for the particular processes that we are about to investigate.

One of the major features of capitalist societies is the use of private property as a storehouse of wealth, which generates a stream of continuing profits and allows further capital accumulation by an individual or institution. In short, in most capitalist societies (and not least in Hong Kong, where property stocks represent around 40 per cent of the local stock-market capitalization), property is seen as one of the most profitable and stable means of capital accumulation – a process that captivates all, from the players within dominant property development companies to the individual home-owner, property speculator or stock-exchange punter. This translates into the search for profit – actual, potential or speculative – which motivates those who accumulate capital, investigate sites and gamble on site development.

Government (the state) may alter the rules or constraints that affect this development process, by mechanisms such as subsidy, taxation, or even direct intervention by carrying some proportion of the direct costs and perhaps by abstracting some of the resulting profit (joint ventures). As long as the state does not hold a dominant interest, development or change will not occur unless the private sector or individual anticipates a profit. The fact that such development processes and decisions relate to sites that happen to be on waterfronts is incidental to the essential features of these processes. It may modify marginally some of the development parameters, rather than bringing about or reflecting any underlying change in the process itself.

Therefore the success or otherwise of individual waterfront projects in Hong Kong (as elsewhere) can be explained by the basic contextual dynamics of the particular property market and its players, and not by any special causal mechanisms associated with waterfronts. Nevertheless, it would be wrong, as we shall see, to conclude that waterfront locations confer no specific benefits. Throughout Hong Kong's history the waterfront has had a particular significance and this will continue to be the case in the twenty-first century. However, bearing these comments in mind, we must first look generally at the dynamics of land and property development in Hong Kong.

LAND DEVELOPMENT IN HONG KONG

Two important concepts underlie the importance of property development in Hong Kong. Reeve (1986) has noted that property is a fundamental institution of social life, and that a society's system of property relations is integral to its economic, political and legal arrangements. Apart from the fact that land sales revenues in Hong Kong have amounted to as much as 45 per cent of total government income in recent boom years, the government has been concerned with major property-related matters, including the actions of property companies and interests, the prices of housing accommodation in a free market, and the provision of a proper legal basis for development of all kinds. Castells *et al.* (1990) have emphasized the importance of property, and particularly the provision of public housing, in the overall macro-economic/political development and control of Hong Kong. It is generally acknowledged that one of the enduring legacies of British colonial rule is the legal administration that regulates business and finance in the territory. Hopefully this will continue after 1997, under the 'one state, two systems' model envisaged in China's Basic Law for the Territory for the next fifty years. The legal framework may be British, but the economic and political life in Hong Kong is local and specific. Hong Kong is 98 per cent Chinese in population and customs, and conducts much of its business in that cultural tradition.

The second concept to remember is that: 'private sector development companies come in a variety of forms and sizes from one-man-bands to multinationals. Their purpose is usually clear-cut: to make a direct financial profit from the process of development' (Cadman and Austin-Crowe 1991: 12). But that simple task is bedevilled by the problems of the property cycle (Fraser 1984: 257–61) which can lead to huge profits, but at the risk of possible bankruptcy. In Hong Kong that process has been accentuated, with three severe property slumps that had major impacts on the local economy between 1960 and 1980 (Ganesan and Tam 1983; Bristow 1984: 267–8; Ganesan 1985).

It is not surprising therefore that the Hong Kong political economy has been dominated by the relationship between landed interests and the government. Traditionally, British colonial territories were governed by an administrative bureaucracy (executive government), advised by members appointed from the 'great and the good' of the local community. This cosy relationship was secure in Hong Kong until the late 1980s, when some directly elected members were brought on to the Territory's Legislative Council (Miners 1991). Prior to this, property companies or leading developers were in effect represented directly on the Legislative Council, and participated directly in government and its many advisory boards and committees. This is illustrated by recent research into the backgrounds of the membership of the Town Planning Board and other development bodies in Hong Kong. The researchers noted that 'membership . . . reveals more than adequate representation of property developers and agents, bankers and other professionals involved in contracting or servicing of property development, not to mention many individuals who are themselves property owners' (Mosher and Taylor 1991: 23). Membership of such bodies confers certain 'useful' benefits, and, despite procedural rules about 'conflict of interest', it is hard to see how 'insider' information could be avoided. This practice was alleged in a court case in 1994 (*South China Morning Post* 1994: 2).

It would be wrong to overemphasize the current importance and influence of this cosy world of big business and government in Hong Kong. The introduction of directly and indirectly elected members on to the Legislative Council in 1991, and more to come in 1995, together with the growth of a better-educated middle class, have placed new political pressures on government which the executive-led system can hardly avoid. These changes are major pointers to a future for Hong Kong governance. The balance of interests in the development story is changing. The principles underlying the processes in the saga which we are about to examine are fundamentally different to those used in the past, and are unique in time and place.

Therefore, we must begin our examination by setting out the general political and economic context for development in Hong Kong, and specifically for the period of the late 1980s and early 1990s during which the airport has been developed. It is well known that 'four factors – the policies of China and of Britain, the attitudes of the local population, and the state of the economy – set the parameters within which the political institutions of Hong Kong must operate' (Miners 1991: xvi). Although today the balance between these factors is changing, all four have always been apparent, even if their definitions have changed over time. Thus, China, Britain, the local population (however defined) and colonial wealth or well-being were the basic policy-making guide for successive Hong Kong governments. This applies as much to land development and the waterfront as to anything else in the territory, and the interplay of these four factors is an important theme in this chapter. In other words, while we have already suggested that mediation between state, property and land interests is essential to a proper understanding of traditional capitalist development processes, in Hong Kong this is bedevilled by two major complications.

First, the make-up of 'government' in Hong Kong has provoked the comment that the state has been infiltrated by the very property and land interests with which it negotiates. Indeed, some would go so far as to say that the state is itself the landed property interest (because almost all land in Hong Kong is available to private owners on a leasehold basis only), reflected in the power of Hong Kong's administrative class (Castells *et al.* 1990: 12–129; Cuthbert 1991: 79). Thus, development decisions by the state become decisions made by and for the landowners' interest.

Nevertheless, in Hong Kong, that interest is subject to a further constraint, which introduces the second complication. Hong Kong is not a wholly independent state. Important mediations have occurred not only between the 'state' and developers but also, in colonial times, between the colonial and home governments, and in more recent times – particularly since the Sino-British Joint Declaration of 1984 – formally between Britain and China. Nowadays, the unfortunate reality is that the larger the infrastructure project, the more complex its international dimension, and the more fraught and important these wider mediations become.

This chapter will show how these mediations have come to affect the implementation of the territory's largest ever infrastructure project – the replacement international airport for Hong Kong (and possibly for South China) at Chek Lap Kok on the north shore of Lantau Island, some 15 kilometres from the urban core. The first runway, the terminal and the infrastructure connecting the airport to the main urban area (Figs 6.1 and 6.2) are due to open one year after Hong Kong becomes a Special Administrative Region of China (1998). This is also the largest waterfront development scheme ever undertaken in Hong Kong, in a long history of such schemes since the colony was founded in the 1840s (Hong Kong Government 1985: endpaper).

To examine this project in depth, it is necessary to explore these four factors first, in terms of the general processes and procedures by which land development in Hong Kong is carried out. In Hong Kong, as elsewhere, reclamation or waterfront development is no more than a detailed or specialized aspect of the general development model.

China needs to be considered first in our discussion. No doubt in the future, the interregnum of little more than 150 years of British rule in Hong Kong will come to be seen as a mere interlude in the thousand-year history of the Chinese mainland. Yet, during that interlude, China has had a benign influence on Hong Kong affairs and this has been greatly strengthened by the terms of the Sino-British Joint Declaration of 1984. As Miners puts it:

> In fact there has been very little change in the way in which decisions are taken. What has altered is the context within which policies are made: China has become much more intrusive, and the Chinese population of Hong Kong have become less apathetic towards public participation.
>
> (Miners 1991: ix)

In fact China has always had an important influence on events in Hong Kong, including land-development matters. For example, the long dispute over the right to demolish and redevelop the infamous Kowloon Walled City led to disputes between Britain and China in the 1930s, and was resolved in the late 1980s with the decision to demolish the Walled City. Equally, the Hong Kong government's decisions to bail out the original Chinese-backed private developers of the Tin Shui Wai new town in 1982, and in 1984 to offer a new site, at a special price, for the National Bank of China in central Hong Kong, reveal a government with a knowing eye on the wider issue of possible Hong Kong/Chinese repercussions.

China has various ways of making its displeasure known, from the pointed criticisms of colonial government policy sometimes uttered by Chinese officials in Hong Kong and elsewhere, to the effects of the 'unfortunate' disruption of the increasingly important ties between Hong Kong and neighbouring regions of China, especially the Shenzhen Special Economic Zone just across the border.

Figure 6.1 Hong Kong: view of the island on which Chek Lap Kok Airport is built
Source: Government of Hong Kong

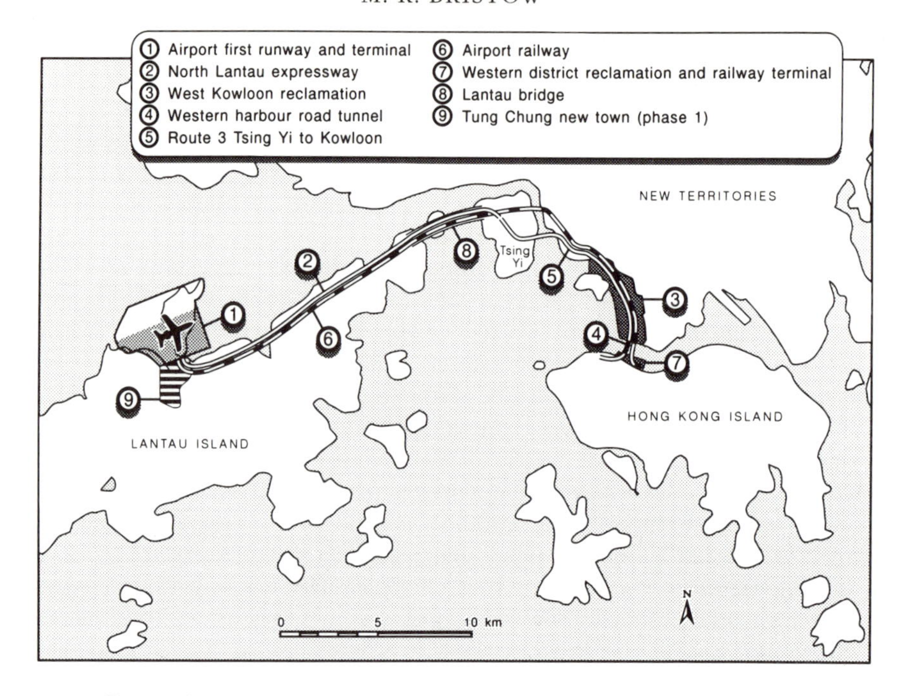

Figure 6.2 Hong Kong: new international airport: nine core schemes

Land development became one of the major issues of the 1984 international negotiations on the future of Hong Kong. One of the specific outcomes of the Sino-British Joint Declaration was the setting up of a joint Land Commission in 1985 to consider development aspects in Hong Kong. The Commission consists of three officials from the Hong Kong side (led by the Secretary for Planning, Environment and Lands), and three from the Chinese side. It oversees the implementation of the terms of the Agreement regarding land leases in the territory, and controls the annual sales of land leases which are nominally limited to 50 ha per year. Half of the proceeds of these sales contribute to a special fund for the benefit of the future government of the Special Administrative Region. When one remembers that in good years up to 45 per cent of the Hong Kong government's annual revenue has come from this source, one can see why all aspects of land development have become so important in Hong Kong, and the significance of China's influence over the process.

Curiously, the workings of the Commission have not yet proved controversial, largely because it is in the interests of both parties to continue existing policies for maximizing revenue. This relationship is only likely to be upset when the government's role as a landowner clashes with its other responsibilities, such as the provision of public facilities (for example in the debates over providing more public support for private-sector, middle-income households in 1991–92), causing policy conflicts over the amounts of land to be released for such purposes.

The pattern of establishing joint bodies has been followed frequently since the Sino-British Joint Declaration of 1984. In particular, as we shall see, the political hiatus in British–Chinese relations that followed the Tiananmen Square incident and the formal initiation of the airport project by the British Hong Kong governor as a morale-raiser for Hong Kong investors inspired formation of the Airport Consultative Committee of 1991. This, however, is a discussion shop with no power.

Direct British influence in Hong Kong is sometimes thought to be greater than it now is – for example in the letting of contracts to British firms in major infrastructure projects such as the new airport, or the various stages of the Mass Transit railway; or in the supposed siphoning-off of colonial funds to London. Britain in fact gave up economic control over Hong Kong in 1974, and political directives from London, apart from those concerning international relations, were rare in Hong Kong even fifty years ago. Now, in the 1990s, with a more politicized Legislative Council in power, such interference is even less likely. In effect, Hong Kong has for many years been internally self-governing, and it is only when international requirements intrude that British involvement becomes essential. As we shall see, this intrusion is one of the more interesting aspects of the new airport story in Hong Kong in the 1990s.

To understand the third area of local interest, the Chinese population, it is necessary to summarize the Hong Kong government's approach to policy-making in land development. For it is only through these mechanisms that local people have any say in local policy (Bristow 1984; Miners 1991). Because the Hong Kong government is not elected, despite the expansion of elected elements in the Legislative Council, Hong Kong's policy-makers must find alternatives to the ballot box to test public acceptance of their proposals. The lack of direct accountability has the advantage that sometimes unpopular but necessary decisions can be easily approved and imposed. Its disadvantage is that government can all too easily 'steamroll' its views, contrary to its claim of openness and consultative policy-making.

How then is policy made in Hong Kong, and how is that process influenced? For those unfamiliar with the Territory, it is perhaps necessary to point out that there is no real equivalent to parliamentary government in Hong Kong. The Legislative Council has formal control over the budget, holds policy debates from time to time, and is becoming adept at questioning the executive, but the power to make policy remains with the bureaucracy led by the colonial Governor and his fourteen appointed advisers in the Executive Council. Broadly, while new policies and legislation frequently originate in the implementing Departments of the government and are endorsed by the senior bureaucracy in the Secretariat (often after ratification by some specialist committee such as the Land Development Policy Committee, first formed in the 1960s), the Governor and his Executive Council have the last word. The Council is similar to the cabinet in parliamentary democracies.

Business people and professionals have direct routes into government policy-making, through individual membership of the government's myriad advisory committees and by direct election of their representatives to the Legislative Council through the functional constituencies, a form of meritocratic double-voting. The general public, however, can operate only through pressure groups or major demonstrations. Two recent examples of this pressure are, first, the controversy over the building of the Daya Bay Nuclear Power Plant in Guangdong Province in China (30 km from urban Hong Kong) which led to major campaigns in the media and on the street and, second, the mass demonstrations in Hong Kong after the Tiananmen Square incident in 1989. While such events are rare, special-interest groups of many kinds have been formed over the years, some permanent, some ephemeral (Miners 1991: 185–95), leading the government to set up consultation mechanisms to absorb potential protesters into its policy-making processes.

The result is a well-practised and generally consistent approach to policy determination and implementation in Hong Kong. Policy is normally prepared and determined through a structure of official committees, either permanent ones, such as the Land Development Policy Committee, or temporary ones, such as those created to undertake particular tasks (for example the steering committees set up to direct and monitor consultant teams working on all aspects of the new airport development). Increasingly, these committees are being opened up to membership by outsiders, or new advisory committees have been formed, such as the new Airport Consultative Committee set up under the Anglo-Chinese Memorandum of Understanding of 1991 (One Country, Two Systems, Economic Research Institute 1992: 117–21).

Government has also begun to publicize its policy proposals in 'Green Papers', such as those on transport, social-service planning, and administrative and electoral reform and in 'White Bills' such as are proposed for the new Planning Ordinance. It has also held formal public consultations, as was done for the new Metroplan covering the main urban area and for the revision of the Territorial Development Strategy in 1993, and at an early stage of the proposed revision of the Town Planning Ordinance. Undoubtedly, the Hong Kong government has tried to become more open, although it may fall short of theoretical democratic control and external constraints may inhibit its plans, especially when international affairs intrude.

The fourth area of concern has always been the economy. Colonial financial secretaries are renowned for demanding balanced budgets, and this principle is upheld in the new Basic Law for the Special Administrative Region:

> Article 107: The Hong Kong Special Administrative Region shall follow the principle of keeping expenditure within the limits of revenues in drawing up its budget, and strive to achieve a fixed balance, avoid deficits and keep the budget commensurate with the growth rate of its gross domestic product.
>
> (One Country, Two Systems, Economic Research Institute 1992: 38)

This is reflected in China's insistence, in mid-1994, that the capital debt for the new airport should be capped at HK$20 billion before it would even consider giving formal approval to the scheme.

At times, this budgetary orthodoxy has caused serious problems for the Hong Kong government. In the 1970s and the early 1980s, for example, public investment in housing and new town development was delayed by recessionary problems in the local economy. In the 1960s, Sir John Cowperthwaite's prolonged opposition to the decision to begin building the new Mass Transit underground railway, partly on the principle of public borrowing, caused considerable problems for the project's planners at the time, and certainly delayed its implementation for several years (Bristow 1984: 247–8). Thus, the bigger the project the more susceptible it may be to general fluctuations in the local economy, which has certainly been true of airport planning in Hong Kong.

A general principle that should always be followed in Hong Kong is never to initiate policies that may be inimical to the overall economic well-being of the territory, regardless of the social or other arguments in their favour. This tenet of official behaviour is as firm now as it was in the past, and continues to influence government policies in all areas of Hong Kong. Prudence remains paramount in economic and budgetary policy in Hong Kong.

THE NEW AIRPORT

Waterfront development in Hong Kong has a long and honourable history. In 1984 it was estimated that some 33 km^2 had been added to Hong Kong's land area by waterfront reclamations since 1841 (Hudson 1979; Hong Kong Government 1985: endpaper). This did not include the major land-use change through the redevelopment of existing waterfront locations (Bristow 1989). The project to relocate and expand the existing international airport is several orders greater than anything so far attempted in any single scheme in Hong Kong. In terms of reclamation alone, the new land created by the scheme is equal to all the reclamation done in Hong Kong from 1945 up to the date when the scheme began. Besides the building of a two-runway airport off the northern coast of Lantau Island, with a 1,248 ha footprint the scheme includes as its 'core' projects: a new railway, connecting expressways, two major bridges, two cross-harbour tunnels, two major harbour reclamations in West Kowloon and Central District, and the first stage of Hong Kong's ninth new town, Tun Chung, adjacent to the new airport. It is not surprising, therefore, that this major set of projects, estimated in March 1991 to involve a total public and private outlay of HK$98.6 billion, has proved controversial and a matter of politics at the highest levels both within and outside Hong Kong. We will now consider this controversy in detail, in order to examine some specific aspects of the political economy of waterfront and land development in Hong Kong. These will enable us to make some broad comments about global land and property development issues.

The history and pedigree of Hong Kong's new airport strategy, first announced by the Governor on 11 October 1989, is long and complex, and has been reported on in substantial detail elsewhere (Pryor 1991). Nevertheless, for our purposes, at least an outline of the saga needs to be given here.

The story began in 1973, when a firm of consultants (Ralph M. Parsons) was asked to undertake a long-term review of aviation development in Hong Kong. In 1975 they recommended the building of a new airport and noted that the best site available was at Chek Lap Kok off the northern coast of Lantau Island (the site now being developed). It was chosen out of the thirteen sites investigated. This proposal resulted in further studies being commissioned in 1978, for the reclamation of the airport site and for land development in North Lantau. The latter was based on studies begun in 1976 on development proposals for the area and the need for new bridge connections from Lantau to the main urban area. A more detailed study on the airport followed in 1980. In the words of government:

> A veritable mountain of deeply researched reports was produced but, taken together, they boiled down to the conclusion that in terms of a new airport site per se, Chek Lap Kok was in all respects a highly suitable choice.
>
> (Pryor 1991: 36)

However, when work was ready to proceed, stringent financial and economic conditions in Hong Kong (due mainly to external pressures) forced the Financial Secretary, Sir John Bremridge, to put the project on hold in February 1983.

Despite the fact that the details of these studies were confidential, there was clearly a general awareness of the government's interest in such matters. It was also clear that firms in the civil-engineering or property development sectors would soon enjoy considerably expanded opportunities – some in the construction of the airport, and others in settlement proposals adjacent to the new airport and in the redevelopment of the old airport site at Kai Tak in the heart of the main urban area.

The development industry, with its representatives in the corridors of power, was thus not ignorant of the plans and as sometimes happens in

Hong Kong, these opportunities were foreseen and seized upon first by one of the territory's major entrepreneurs. In November 1986, Gordon Wu of Hopewell Holdings proposed three schemes to the government for joint airport and seaport development in the Western Harbour (*Far Eastern Economic Review* 1987a and 1987b). Wu's proposals were a response to increasing government interest in joint venture and large private infrastructure development (Bristow 1989: 215). They were also an attempt to utilize the specific demands for port and airport expansion in Hong Kong as a major private development opportunity. The port expansion requirements, as reported earlier to government (Marine Department 1986), offered options for extending the Hong Kong Container Port, begun in the 1960s, beyond its site at Kwai Chung in the Western Harbour, and related these proposed changes to other aspects of port development in Hong Kong.

Government, however, was unwilling to accept Wu's proposals directly, because, to a degree, they challenged government thinking on harbour and airport development. Perhaps too, despite its wish for a more equal partnership between the public and private sectors in major infrastructure development, it did not want to see such a big scheme being driven openly by the Hong Kong development industry. The private-sector initiative did, however, result in a further investigation. This was partly a response to updated traffic forecasts and partly to reconfirm and consolidate ideas about where the port development and new airport might be built on an integrated basis.

The response to Wu's proposals, made by the Executive Council in 1987, had three components, that:

> planning for the development of Chek Lap Kok and possible alternative sites should proceed; endeavours should be made to ensure without commitment, that any consultative studies undertaken by the Hopewell Group should be relevant to government's own plans; and at an appropriate time, it would be necessary to involve the PRC government [China] in taking the plans forward so that any development would be consistent with their plans and, furthermore, with a view to avoiding saddling the Hong Kong SAR government with an unacceptable burden of debt.
>
> (Pryor 1991: 48)

In the event, the government returned to its trusted band of existing consultants for advice on what now became known as the PADS (Port and Airport Development Strategy) Study. This produced a Final Report in December 1989 (Mott, MacDonald, Hong Kong Ltd *et al.* 1989).

Three main options were finalized and evaluated: to retain the existing airport at Kai Tak; to build a new airport at Chek Lap Kok as previously agreed; or to locate the new airport on a site between Cheung Chau and Lamma, two islands in the Western Harbour. More studies were conducted to propose a strategy for port development beyond the two new container terminals (Eight and Nine), to which the government was already committed (to be constructed in the mid-1990s on Stonecutters and Tsing Yi Islands respectively, close to the existing container port), and to look at how the whole undertaking might be financed (Pryor 1991: 55–6). The re-evaluation of the proposed sites for the airport was in turn analysed by a Dutch consultancy (Netherlands Airport Consultants 1989).

After this round of investigations, a series of presentations was made to the Executive Council of Hong Kong in September and October 1989, and the Governor announced in his annual address to the Legislative Council, on 11 October 1989, that the new airport at Chek Lap Kok would go ahead. Since then legislation setting up the Provisional Airport Authority has been passed (1990) and a series of consultant, contractor, and internal committee structures to implement the works appointed and approved (important official ones are the Port Development Board and the New Airport Projects Co-ordination Office – both in the Government Secretariat) (Pryor 1991: 66–108).

This study is not, however, a report on how these projects are to be implemented, but on why and how their implementation so far has been influenced and affected by outside factors. This is where intelligent speculation is needed, because so much takes place behind closed doors – an echo of Kirk's (1980) earlier observation. But even though the projects are contemporary and not yet complete, and many discussions remain confidential, there is already enough information in the public domain to enable us to make some well-informed guesses, some of which are highly relevant to processes and procedures elsewhere. We shall now turn to the analysis of the general significance of the Hong Kong scheme.

EVALUATION

Based on our review of policy-making in Hong Kong, we can see here an internal, committee-led decision process, similar to those followed elsewhere within government, although this one was made particularly complex by the logistics of modern airport development. In this case, however, we need to consider why the decision process has been complicated. It will be useful to recall the four influences on Hong Kong governmental actions: China, Britain, local pressures and the economy.

That China should come first is entirely appropriate, for in the 1990s the new airport became a political matter involving China and its international relationships at the highest level. Hong Kong's sensitivities towards China were not new. We have already mentioned some earlier infrastructural development decisions which were clearly influenced by mainland considerations. The airport development in Hong Kong magnified these considerations. A replacement airport in Hong Kong was considered as early as the 1940s and the suggested site was in Deep Bay (Nim Wan), a sea inlet in the North West New Territories near the Chinese border. Then, and in 1982 when a similar idea was investigated, the problem of overflying Chinese airspace, thereby potentially coming under Chinese control, was a primary reason for the site's rejection (Pryor 1991: 48). But it is equally clear from later government statements, that one reason for halting the original airport replacement scheme in 1983 was the state of bilateral relations between Britain and China prior to the Sino-British Joint Declaration of the following year (*Hong Kong Hansard* 1990: 541). Much more important in the present case, however, was the timing and manner of the Governor's announcement of approval for the project in October 1989.

The year 1989 was momentous for China and Hong Kong. In China the pro-democracy and anti-corruption movement came to a climax, which in July led to the bloody repression in Tiananmen Square. In Hong Kong there was a wave of support for more democracy in China, and mass protests at the actions of the Chinese government. The events in China, and reactions in Hong Kong, upset the delicate negotiations over the transitional arrangements for Hong Kong's hand-over to China in 1997, and in particular it raised doubts in Hong Kong about how China would behave there after the take-over. This affected the final drafting, on the Chinese side, of the new Chinese Basic Law for Hong Kong after 1997 (Chan 1991: 17–29), bringing in new amendments to reinforce political safeguards for Beijing. However, it also led to more strident indirect interventions by China in Hong Kong affairs from 1991 until the time of writing in 1996. These interventions concerned particularly the changes in elections to Hong Kong bodies, and budgetary and investment matters focusing on the proposed new airport and its associated developments. China skilfully procrastinated on the latter (indirectly threatening the Hong Kong economy) in an attempt to effect favourable decisions on the former (its future political control).

Initially, in 1989–90, after the project was announced in Hong Kong, four rounds of expert consultations on the new airport proposals were held with Chinese officials, which were eventually resolved by the direct intervention of the British Foreign Office and the British Prime Minister. It was said (in the Anglo-Chinese confrontations over increased democracy in Hong Kong from 1992 to 1994) that John Major, the British Prime Minister, was furious at being forced to go to Beijing in September 1991 to sign the resulting joint 'Memorandum of Understanding' (Mosher 1991: 10–12).

The Memorandum (One Country, Two Systems, Economic Research Institute 1992) agreed the nine core projects associated with the scheme, but more importantly it set parameters for China to exploit later in its interventions in Hong Kong affairs. This in effect tied the airport development process to the entanglements then beginning over the political development and institutional changes in Hong Kong in the run-up to 1997.

> Although many questions and criticisms were raised against the airport and related projects, for example, whether Chek Lap Kok was the best site, most were red herrings. From the start there were only two real issues: the financing of the project, and China's right to be 'consulted' on all major Hong Kong matters straddling 1997.
>
> (Ng 1991: 80)

Even though in November 1994 the matter of finance finally had been resolved in principle, in a way both matters seemed to remain as probable foci for ongoing disputes between Britain and China. On far too many occasions, airport matters were used as a lever in Anglo-Chinese political relations, particularly in 1993–94 when these were especially strained by increased democratic development in Hong Kong.

By consultation, China clearly means that its views must be considered before the Hong Kong government makes any decisions. According to Ng, 'for China, it goes further. "Consultation" meant that where there is a difference of opinion, there should be negotiations between the British and Chinese governments, until agreement is finally reached, and that no action is to be taken until there is agreement' (Ng 1991: 80). The difficulty for the British and Hong Kong governments and for the airport scheme is that this view challenges the competency and power of both governments to control Hong Kong affairs until 1997, and allows China to use delays or rejection of decisions on a project (all of which cost money) as levers on other aspects of Anglo-Chinese relations. Following the 1991 Memorandum of Understanding, China did not fail to use such weapons, and the Hong Kong Airport scheme and its financing became a matter of international diplomacy.

Thus, in its most serious confrontations until late 1992, China threatened to refuse to recognize contracts in Hong Kong that continued after 1997, unless previously accepted by its government alone. This would affect the construction and financing of the new airport and extensions to the container port among others. China also threatened to reject new stages in the scheme, some involving important design changes such as re-siting the terminal complex or re-aligning the first runway. Because such decisions had to be taken by the Hong Kong government alone, Chinese officials now felt able to challenge the costs involved. By late 1992 they had threatened to refuse formally to agree to the letting of necessary contracts, forcing the Hong Kong government to proceed unilaterally or risk postponement and extra costs. The new airport became an issue of full confrontation between Britain, China and Hong Kong, and was used as a pawn in the diplomatic battle over the democratization of Hong Kong in the period prior to 1997. This issue was not formally resolved until the signing of yet another Sino-British agreement on 4 November 1994, on the financing of the scheme.

Turning to Britain's role, Castells' review of governance in Hong Kong recalls the continuing importance of the Governor in Hong Kong affairs (Castells *et al.* 1990: 119–32). British influence on the new airport has been benign, and focused on the actions of the new Governor, Chris Patten, appointed in 1992. Unlike his predecessors, Patten is a politician, not a diplomat. Even more noticeable, in recent years, is that increasingly matters important to Hong Kong have been negotiated between Britain and China without direct Hong Kong representation. An early example of this was China's refusal to accept a Hong Kong passport-holder as one of the British negotiating team for the original Sino-British negotiations of 1982–84 about the future of Hong Kong after 1997. Much of the supposed offence that led to the Sino-British confrontation in late 1992 was caused by the new Governor's tendency to act as if in sole control, doubtless with approval from London, on democratization, investment and implementation of the airport scheme. China found this unacceptable in a supposedly subservient power, and dangerous as a precedent for Hong Kong's actions after 1997.

Britain has been criticized for allegedly exercising favouritism in the award of contracts for the airport, even though the project is open to competitive tendering world-wide. Some commentators in Hong Kong say that in negotiations Britain has tended to support its own relationship with China, rather than support Hong Kong in a confrontation with China. Nevertheless, ever since the new airport scheme was formally approved in 1989, British policy has consistently supported the scheme – partly as a means of generating economic and political 'faith' in Hong Kong's long-term future and partly to show outsiders, particularly China, that the Hong Kong government will retain full control of its affairs up to the deadline of 30 June 1997. Britain's attitude, like China's, has been determined by the image of its international stance.

In relation to Hong Kong, discussions on and investigations into the new airport proposals were initially secretive, although there were some people 'in the know' and the debates and questions of the Legislative Council shed occasional shafts of light on the project. The announcement by the Governor, Sir David Wilson, of government's decision to proceed in October 1989 therefore marked a beginning rather than the end of a process of public discussion on the topic. The preliminary rounds of wide private consultation – another procedure typical of Hong Kong – were, in the event, insufficient to remove public doubts, which became apparent once the proposals were publicized. It is fair to agree with Ng that, initially:

> There was no room for public participation on the part of various interested parties [stakeholders] at critical stages in the development of the plan. Studies done on the airport were not made public and the people were left in the dark about the progress of these projects.
> (Ng 1993: 302)

Indeed, government's sensitivity was soon shown by its speedy decision to publish its consultants' final report and recommendations on the scheme (Mott, MacDonald, Hong Kong Ltd *et al.* 1989) in 1991, for sale to the public, as a part of its response to its critics and as an attempt to justify its position.

In response, there were letters to the newspapers, expressing concern at the cost of the scheme, and two private professional teams devised and presented alternative siting strategies for the airport. These were additional to Gordon Wu's scheme, which had prompted the government's energetic defence of its original proposals. Dr L. H. Wang's group suggested (Fig. 6.3) a site in Nim Wan (Deep Bay). The other sites were Clearwater Bay, or Tolo Harbour in the east of the territory (Study Group for Infrastructure Development 1991). However, both were rejected by government on the grounds of inadequate evaluation and analysis, insufficient allowance for necessary transportation links or incompatibility with the government's views on overall planning strategies for the Territory as a whole. Effectively rubbishing its critics in a press release, the government said of the second study:

> Whilst it fully appreciates that the study has been carried out with good intention within the limits of resources, expertise and time available . . . the government considers the report lacking in detailed technical analysis in respect of aeronautics, traffic projections, environmental impact, and so on, which are required to enable a scientific and sensible comparative assessment to be made.
>
> (Government Information Service 1991: 13)

In short, despite severe criticism from many of the directly elected members of the Legislative Council in its airport debate of November 1991, the government was able to see off its public critics by claiming superior knowledge – not difficult given the resources it could use to gain that expertise.

More interesting perhaps, in terms of participation, is the role of the special Consultative Committee, set up under the 1991 Memorandum of Understanding between Britain and China. The relevant sections of that Memorandum read as follows:

> (F)(iii) The Hong Kong Government will set up a Consultative Committee on the new airport and related projects. The Committee may discuss any relevant matter but will have no decision-making power. It should not delay the progress of the projects.
> (iv) The Hong Kong Government will inform the Chinese side of the members of the Airport Authority and Consultative Committee whom it is proposed to appoint, and will be willing to listen to any views that the Chinese side might have, before deciding on the appointments . . .
> (One Country, Two Systems, Economic Research Institute 1992: 119–20)

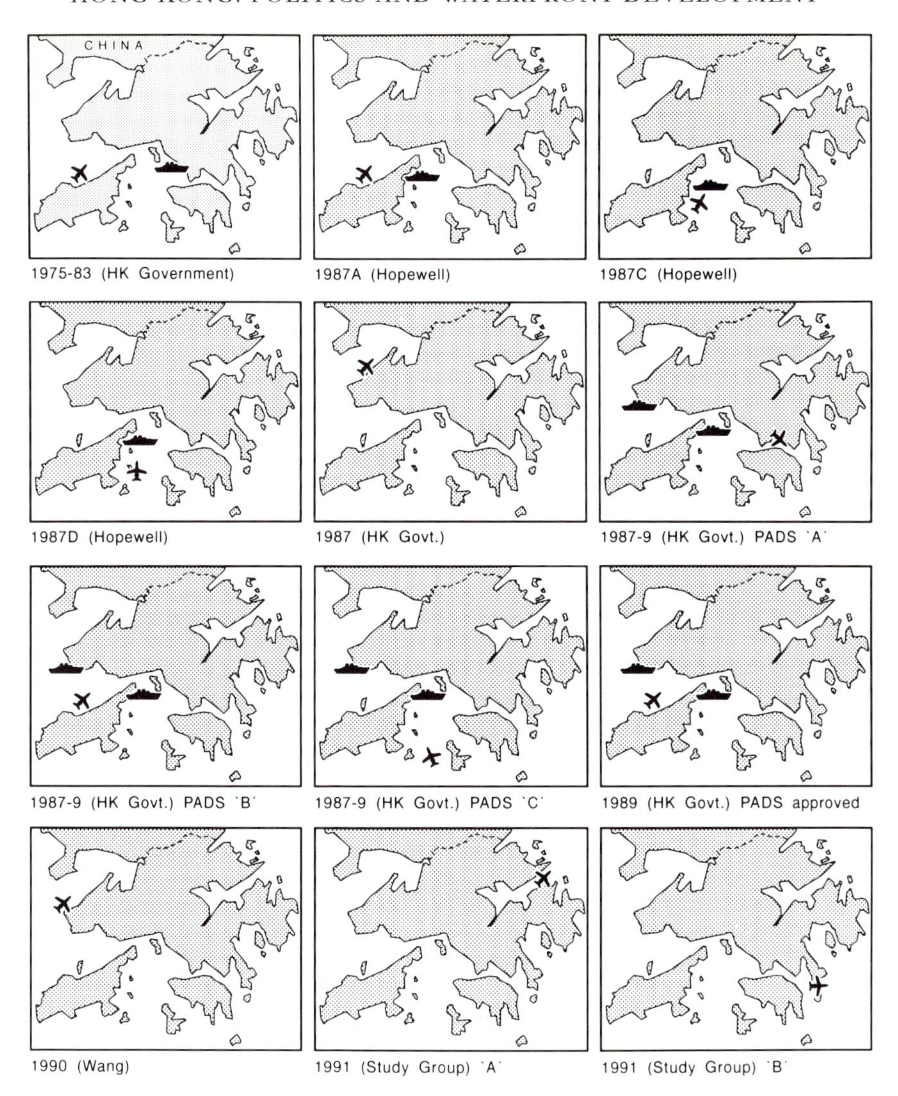

Figure 6.3 Hong Kong: sites investigated for the airport during the period 1975–1991

As with other consultative groups set up recently under Sino-British auspices, its membership is large (fifty people) and designed to contain a wide cross-section of the community. Although the Chief Secretary of the Hong Kong government said that it would provide a valuable forum for collecting community views on the projects and for the government to explain the projects to the public, its constitution and lack of power has reduced it to little more than a faction-ridden talking-shop, looking over its shoulder at its various political sponsors. Nevertheless, it has met quite frequently (ten meetings in 1992) and spawned four subcommittees relating to the airport and associated land developments – traffic and transport, finance and planning, and the environment. It has often been used as a forum for setting out more detailed planning proposals for each of the nine core projects as they have been developed and brought forward since 1989. 'Real' decisions, however, have been formulated or argued elsewhere, either in government or in the Sino-British Joint Liaison Group set up under the 1984 agreement. The decisions are then presented to the Consultative Committee or the public as a series of prejudged and predetermined decisions for ratification.

This is symptomatic of much that government does in Hong Kong. It is not a new policy. As the Financial Secretary, in replying to the airport debate in the Hong Kong Legislative Council on 21 November 1990, put it:

> Given the importance of the PADS projects to our future, the Government fully appreciates the wish of and the need for the public in general, and this Council in particular, to be kept informed. . . . Sir, one of Hong Kong's great attractions is that it is possible to get things done here. Plans, once they are agreed, are implemented quickly. As Sir David Ford indicated . . . to hold exhaustive public enquiries, which is the route followed in some places, is not always the swiftest way of achieving results. It has its cost – the cost is delay after delay after delay. That is not Hong Kong's way of doing things. The Hong Kong style is to take a vision of the future, to decide boldly what needs to be done to achieve that vision, and then to get on with it.
>
> (*Hong Kong Hansard* 1990: 542)

But it is a vision that may not be shared by all in Hong Kong. Openness seems to mean information only, not the right to discuss or adjudicate on the decisions being made. This procedural method is beginning to be challenged in the newly politicized Legislative Council.

CONCLUSIONS

The development of the new airport and its associated waterfront schemes in Hong Kong presents an interesting contrast to controversies stirred up elsewhere by major airport schemes. It does not compare, for example, with the saga surrounding the Third London Airport (Hall 1980: 15–55; Buchanan 1981), or indeed with the opposition to airport expansion proposals in other world cities such as Frankfurt or Tokyo. Why should this be, or is the project in some way different?

Certainly, in terms of infrastructure, Hong Kong's airport project is comparable with any other major airport scheme in the world, and is bigger than most. Therefore it has significant environmental and social/community impacts. Yet reactions so far in Hong Kong seem different to those elsewhere. Maybe this can be explained by what Cuthbert has said in another context: 'in Hong Kong we may witness a totally unique example of urban development undisturbed by social conscience' (Cuthbert 1991: 78). He points out that:

> Hong Kong has never possessed a democratic system of politics, and the local people have never had the opportunity to vote for their own leaders. Government is composed of two bodies, the executive and legislative councils whose structure is such that the Governor has total control over policy. But in reality hegemonic control is exercised by the private sector since it provides the government with its essential tax revenue from land leases, taxes on profits and its control over public utilities. Consequently, private sector interest massively penetrates both government councils, where both appointed and elected members represent all major companies through directorships.
>
> (Cuthbert 1991: 79)

Thus, for an explanation of the how and why of waterfront development in Hong Kong, and of the airport in particular, we must return to the specifics of the contemporary local political economy: who has power, and how and why it is wielded. While some might accuse Cuthbert of overstating his case, it is difficult to refute his explanation. State policy and private-sector interest (of capital) in Hong Kong have combined in the airport scheme to take on all comers, including China. As a more obvious example of naked self-interest than is usual in the murky world of property development, planning and infrastructure development, it provides useful insights into property and waterfront development elsewhere. As a landowner and holder of the key interests of Hong Kong in economic development, the state now also reflects the developers' interests in economic growth and stability. In short, the interests have become one and the same, reinforced by the intermingling of state and capital in the body politic of Hong Kong.

The airport scheme in Hong Kong is going ahead because it is in the government's interest (and Britain's interest) to be seen to be giving massive backing to Hong Kong development (neither government wants to have to deal with the collapse of the Hong Kong economy and the rise of a serious refugee problem in the run-up to 1997). The private sector sees greater and richer pickings to be made in the fields of real estate and contracting, should the scheme be implemented. Against such dominant interests, any community interest is immaterial, marginalized and not influential. The scales are not equally balanced in Hong Kong for:

> as many people have rightly argued, the political intricacies in the territory have rendered the development of a genuine strategic planning process a remote ideal. An open planning system vital for the implication of a strategic planning process can only exist within a democratic political structure.
>
> (Ng 1993: 308)

Once again we must consider the mediation of interests (Healey *et al.* 1988). To be meaningful this mediation has to be between interests with substantial, if not equal, power. The community in Hong Kong does not have that level of power, and, if China has its way, may never have it. This brings us back to reconsidering those parallel arguments of the 1990s about democracy in Hong Kong: what it might mean and how it might be implemented. In the long term, the answers are as important for planning and development as for other fields of public policy in the territory.

Whether the lessons of this Hong Kong debate about a particular infrastructure project can be applied elsewhere is uncertain, for, just as success in Hong Kong depends on the territory's unique political economy, so too does it reflect values and ethics that may not be universally acceptable or exportable. Nevertheless, its procedures do point to one way in which successful waterfront projects on a massive scale can be achieved rapidly and within budget. In this, if nothing else, there are indeed lessons to be learnt and questions to be answered about appropriate methodologies for development for the global community.

BIBLIOGRAPHY

Breen, A. and Rigby, D. (1994) *Waterfronts: Cities Reclaim Their Edge*, New York: McGraw-Hill.

Bristow, R. (1984) *Land-use Planning in Hong Kong – History, Policies and Procedures*, Hong Kong: Oxford University Press.

—— (1989) *The Hong Kong New Towns – A Selective Review*, Hong Kong: Oxford University Press.

Buchanan, C. (1981) *No Way to the Airport*, Harlow: Longman.

Cadman, D. and Austin-Crowe, C. (1991) *Property Development*, London: E. and F. N. Spon.

Castells, M., Goh, L. and Kwok, R. Y.-W. (1990) *The Shek Kip Mei Syndrome – Economic Development and Public Housing in Hong Kong and Singapore*, London: Pion.

Chan, M. K. (1991) 'Democracy derailed – realpolitik in the making of the Hong Kong basic law, 1985–90', in M. K. Chan and D. J. Clarke, *The Hong Kong Basic Law – Blueprint for 'Stability and Prosperity' under Chinese Sovereignty?*, Hong Kong: Hong Kong University Press: pp. 3–35.

Cuthbert, A. R. (1991) 'The water margin – urban design, reclamation and development in Hong Kong', in *International Centre, Cities on Water, Waterfronts – A New Urban Frontier: Papers and Abstracts*, Venice: International Centre, Cities on Water: pp. 78–81.

Far Eastern Economic Review (1987a) 'Dredging up the future', *Far Eastern Economic Review* 135(5), 29 January: 46–7.

—— (1987b) 'The Hong Kong quartet', *Far Eastern Economic Review* 135(12), 19 March: 114–16.

Fraser, W. D. (1984) *Principles of Property Investment and Pricing*, London: Macmillan.

Ganesan, S. (1985) *Property Cycles – Theoretical Aspects and Empirical Observations on Hong Kong*, Hong Kong: Centre of Urban Studies and Urban Planning, University of Hong Kong, Working Paper no. 11.

Ganesan, S. and Tam, I. (1983) 'The crisis in the property sector of Hong Kong and the future', in Appointments Board, *Architecture, Building, Urban Design and Urban Planning in Hong Kong*, Hong Kong: University of Hong Kong.

Government Information Service (1991) 'Lantau the best overall site for PADS', in *Government Daily Press Releases*, Hong Kong, Wednesday, 3 July: 13–15.

Hall, P. (1980) *Great Planning Disasters*, London: Weidenfeld and Nicolson.

Healey, P., McNamara, P., Elson, M. and Doak, A. (1988) *Land Use Planning and the Mediation of Urban Change: The British Planning System in Practice*, Cambridge: Cambridge University Press.

Hong Kong Government (1985) *Hong Kong 1985 – A Review of 1984*, Hong Kong: Government Printer, endpapers.

Hong Kong Hansard (1990) 'Speech by the Financial Secretary, Sir Piers Jacob, 21 November 1990', *Hong Kong Hansard*, Hong Kong: Government Printer: pp. 541–3.

Hong Kong Institute of Engineers (1990) *Airports into the 21st Century*, Hong Kong: Hong Kong Institute of Engineers, Conference Proceedings.

Hudson, B. J. (1979) 'Coastal land reclamation with special reference to Hong Kong', *Reclamation Review* 2 (1): 3–16.

Kirk, G. (1980) *Urban Planning in a Capitalist Society*, London: Croom-Helm.

Kymlicka, W. (1990) *Contemporary Political Philosophy – An Introduction*, Oxford: Clarendon Press.

Low, N. (1991) *Planning, Politics and the State – Political Foundations of Planning Thought*, London: Unwin Hyman.

Marine Department (1986) *Port Development Strategy Study – Final Report*, Hong Kong: Marine Department.

Miners, N. (1991) *The Government and Politics of Hong Kong*, Hong Kong: Oxford University Press.

Mosher, S. (1991) 'Creeping intervention', *Far Eastern Economic Review* 153(29), 18 July: 10–11.

Mosher, S. and Taylor, M. (1991) 'Undeclared interests', *Far Eastern Economic Review* 154(45), 7 November: 22–5.

Mott, MacDonald, Hong Kong Ltd, Shankland Cox, Maunsell Consultants Asia Ltd, Wilbur Smith Associates, and Coopers and Lybrand Associates (1989) *Port and Airport Development Strategy – Final Report*, Hong Kong: Government Secretariat – Land and Works Branch.

Netherlands Airport Consultants (1989) *Alternative Replacement Airport Sites Consultancy – Final Report*, Amsterdam: Netherlands Airport Consultants BV.

Ng, M. K. (1991) 'The implementation of the Sino-British Joint Declaration', in Y. W. Sung and M. K. Lee, *The Other Hong Kong Report 1991*, Hong Kong: Chinese University Press: pp. 77–88.

—— (1992) 'The politics of planning and regional development: a case study of the container port and airport development in Hong Kong', University of California Los Angeles, unpublished Ph.D. thesis.

—— (1993) 'Strategic planning in Hong Kong: lessons from TDS (Territorial Development Strategy) and PADS (Port and Airport Development Strategy)', *Town Planning Review* 64 (3), July: 287–311.

One Country, Two Systems, Economic Research Institute (1992) 'The Basic Law of the Hong Kong Special Administrative Region of the People's Republic of China': with Sino-British Joint Declaration; 'Memorandum of Understanding Concerning the Construction of the New Airport in Hong Kong and Related Questions', Hong Kong: One Country, Two Systems, Economic Research Institute.

Pinder, D. A., Hoyle, B. S. and Husain, M. S. (1988) 'Retreat, redundancy and revitalisation – forces, trends and a research agenda', in B. S. Hoyle, D. A. Pinder and M. S. Husain, *Revitalising the Waterfront – International Dimensions of Dockland Development*, London: Belhaven.

Pryor, E. G. (1991) *Hong Kong's Port and Airport Development Strategy*, Hong Kong: Government Information Services.

Pun, P. K. S. (1991) 'Planning for the waterfronts in Hong Kong', in International Centre, Cities on Water, *Waterfronts – A New Urban Frontier: Papers and Abstracts*, Venice: International Centre Cities on Water: pp. 23–5, 31–41 and 60–71.

Reeve, A. (1986) *Property – Issues in Political Theory*, London: Macmillan.

South China Morning Post (1994) 'Confusion over board declaration', Hong Kong: 9 August: 2.

Study Group for Infrastructure Development (1991) *Alternative Sites for Hong Kong's Replacement Airport – A Prefeasibility Study: Executive Summary*, Hong Kong: Study Group for Infrastructure Development (Main Report in Chinese).

7

TOKYO, OSAKA AND KOBE: ISLAND CITY PARADISE?

Yoshimitsu Shiozaki and Patrick Malone

INTRODUCTION

This chapter presents a critical examination of three new marine cities built on artificial islands in Japan. The three island cities are located in the bays of Kobe, Osaka and Tokyo. By examining these three projects, we intend to counter the arguments in favour of artificial islands and large marine developments, and to determine why so many of these developments have been built in Japan.

Land reclamation has been used for a number of years in the development of Japan's ports. In the period between the 1950s and late 1970s, the expansion of Japanese industrial production and rising levels of exports generated a need for larger and more efficient port facilities. Initially, in order to promote efficiency, harbour areas were legally confined to port-related forms of development. Gradually, however, the emphasis on the development of port facilities has eased in favour of mixed-use projects.

Although recent projects may embody new port facilities, there is a trend towards developments that combine office, commercial, housing, parks and other forms of land use. This shift in the balance between port facilities and other functions means that marine developments have a new role in urban and economic development. In effect, large marine projects increasingly provide extensions to parent cities rather than simply to their ports.

While the approach to marine developments in Japan is in a state of perpetual evolution, land reclamation is a major factor in the development of waterfront cities. To some extent, the popularity of land reclamation reflects costs and problems associated with the development of existing land. In some cities, reclamation is as significant as the development, or redevelopment, of existing land.[1] It is important, however, to distinguish between:

(a) developments that extend existing urban or port areas using reclaimed land
(b) artificial islands
(c) projects that have the status of island cities

To date, four island cities have been built in Japan. In addition to the three developments covered in this chapter, Kobe has a second island city (Rokko Island City) (Fig. 7.1).

In some respects, these island cities resemble inland new towns. However, while Japan's inland new towns are based mainly around the provision of housing, island cities may incorporate port facilities and housing may be a secondary element.[2] Moreover, inland and marine developments raise different questions. In island cities, the degree of separation from the mainland presents specific problems and may colour issues relating to housing and living conditions. For the residents of mainland and new marine developments, the question of 'urban quality' is related to the nature of transport networks, housing, public amenities and welfare facilities. Because they are isolated from the mainland, however, problems associated with those factors can be more acute in marine developments. It is significant, therefore, that the centres of Yasio Park Town and Nanko Port Town are 9.5 and 10 km respectively from the centres of their parent cities of Tokyo and Osaka, while the centre of Port Island is 2.5 km from the centre of Kobe.

Marine developments also raise specific issues relating to their environmental impact on waterfront and urban areas. Thus, mainland residents who have large waterfront projects on their doorstep may be concerned about the environmental and other effects of marine developments on the parent city. As recent experience in Kobe has shown, marine developments and artificial island cities also raise very serious questions in terms of their vulnerability to earthquakes. Here, the question is whether artificial islands are more vulnerable than comparable inland developments to the effects of earthquakes, particularly in terms of the damaging effects of shock on islands, buildings and transport links to the mainland.

This chapter focuses on the issues raised directly by large marine developments, and specifically by artificial island cities. We will argue that there may be doubts concerning the quality of Japan's first three marine cities at Tokyo, Osaka and Kobe. These doubts relate to planning, environmental factors and the capacity of these island cities to provide satisfactory living conditions in the face of social needs.

This chapter also suggests that many of Japan's large marine developments are derived from quantitative planning, economic objectives and the ambitions of urban governments to expand urban economies. The reasons commonly given for the reclamation of land and large marine developments reflect the need for new port facilities, more efficient systems for the distribution of goods, and increased economic activity and consumption. As stated above, the demand for new land for offices, housing and other functions has also emerged as a major factor over the past decade. Housing and other land uses may be secondary in those projects that were developed to meet the demands for new port facilities; however, the tendency towards mixed-use marine projects was boosted by the economic boom of the 1980s and by the drift towards deregulation and privatization in Japan (see Chapter eight).

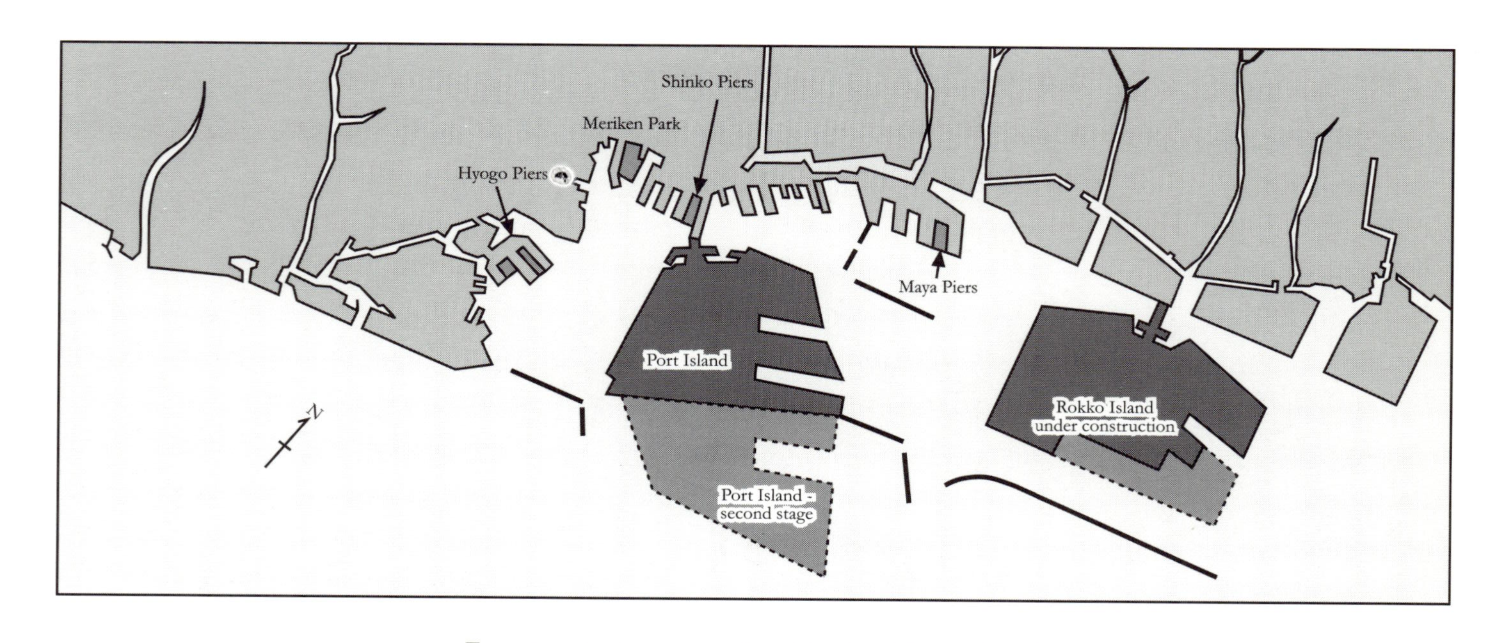

Figure 7.1 Kobe: map of Port Island and Rokko Island
Source: Port Authority, Kobe

City governments are the major powers behind marine developments although, clearly, these projects provide substantial profits for the construction industry and other forms of capital. Despite Japan's shift towards deregulation, urban governments retain a powerful role in urban development. They continue to promote marine projects in terms of urban and regional economic growth and the expansion and improvement of the ports which underpin growth. It is also claimed that, as extensions to the city, large marine developments may help to solve existing urban problems by alleviating congestion in urban cores and easing pressures for office space, housing, recreational facilities and open space.

Clearly, Japan's marine developments feed ambitions for urban and economic growth. They also generate substantial profits for development and other interests. In the final analysis, however, the need for large marine developments and island cities should be subject to question; particularly because, as they move away from the provision of port facilities, Japan's marine developments raise more questions about planning objectives and social needs.

THREE ISLAND CITIES

The three island cities examined in this chapter (Port Island, Kobe; Nanko Port Town, Osaka; and Yasio Park Town in Tokyo) are Japan's first artificial mixed-use island cities although, as stated above, Kobe has now a second island city.

Japan's artificial islands are built mainly from soil taken from the mainland, mountain areas or the sea-bed, although refuse and waste from building and subway construction are also used. Port Island, for example, was built of sandstone debris which was carried to the water's edge on a 14-km-long underground conveyor belt, and from there by barge to the site. Typically, the material used to create islands is carried by barges and dumped between retaining walls to a height of 15–20 metres.

All three projects were pioneered and planned mainly by city or local authorities, although the national Housing and Urban Development Corporation has a major role in terms of housing. Nanko Port, for example, was developed by a special department of the Osaka Port and Harbour Bureau which was involved initially with the reclamation of the land for the project. This department (known as the Nanko Development Division) and an allied project team were responsible for the planning, promotion and management of the project. During the development process, the Bureau invited construction companies and developers to respond to its master plan; it approved detailed plans and had overall charge of the project. It also played a key role in the provision of transport and other infrastructure, parks and public facilities. Economically, it was empowered to sell land, to recover the costs of reclamation and to fund the public elements of the project. Thus, the development system for these island city projects is characterized by the involvement of powerful public development authorities which deal directly with construction and development interests over a relatively fixed master plan. This system leaves less scope for private development and property capital than many development systems operating on waterfronts outside Japan.[3]

A comparative analysis shows that Osaka's Nanko Port Town is the largest of the island cities (937 ha), followed by Tokyo's Yasio Park Town (679 ha) and Port Island (436 ha)(Table 7.1). Nanko Port Town is also the largest project in terms of its target population (40,000). Tokyo's Yasio Park Town and Port Island, although different in size, have roughly comparable planned populations (18,000 and 20,000 respectively).

The land-use structures of the three cities can be compared in terms of the nature and hierarchy of the functions that they accommodate (Figs 7.2, 7.3 and 7.4). In each, port and other commercial facilities are the dominant land uses, followed by public open space and housing. The housing element

Table 7.1 Tokyo, Osaka and Kobe: summary data on three island cities

Island city	*Yasio Park Town*	*Nanko Port Town*	*Port Island*
Parent city	*Tokyo*	*Osaka*	*Kobe*
Location	Shinagawa Ward	Suminoe Ward	Chuou Ward
Year started	1972	1975	1966
Total area (ha)	679.8	936.8	436.0
Target population	18,000	40,000	20,000
Area of residential zone (ha)	41.0	100.0	23.0
Target no. of households	5,270	10,000	6,588
General land use, site area (ha)			
Total area	679.8	936.8	436.0
Housing	26.0 (3.8%)	48.7 (5.2%)	22.3 (5.1%)
Parks	92.0 (13.5%)	81.1 (8.7%)	24.0 (5.5%)
Port facilities	436.0 (64.6%)	498.8 (53.2%)	233.0 (53.4%)
Commercial/other	125.8 (18.1%)	308.2 (32.9%)	136.3 (26.0%)
Public waterfront	4.0 km	2.0 km	0.5 km

Sources: Tokyo Metropolitan Government Housing Bureau; Housing and Urban Development Corporation (1980) *Master Plan of Shinagawa Yasio Housing Estate*; Port and Harbour Bureau City of Osaka (1983) *Osaka Nanko*; Kobe City Development Bureau (1979) *Port Island*
Notes: Actual populations recorded are: Yasio Park Town 16,899 (1989); Nanko Port Town: 28,600 (1985); Port Island 14,000 (1984). Start dates shown refer to the date that the development began on site; however, projects may have a long history in terms of previous development proposals or land-reclamation processes. Housing areas include areas under houses, gardens, pathways, etc. (but not trunk roads). Areas shown for residential zones include retail, commercial and social facilities.

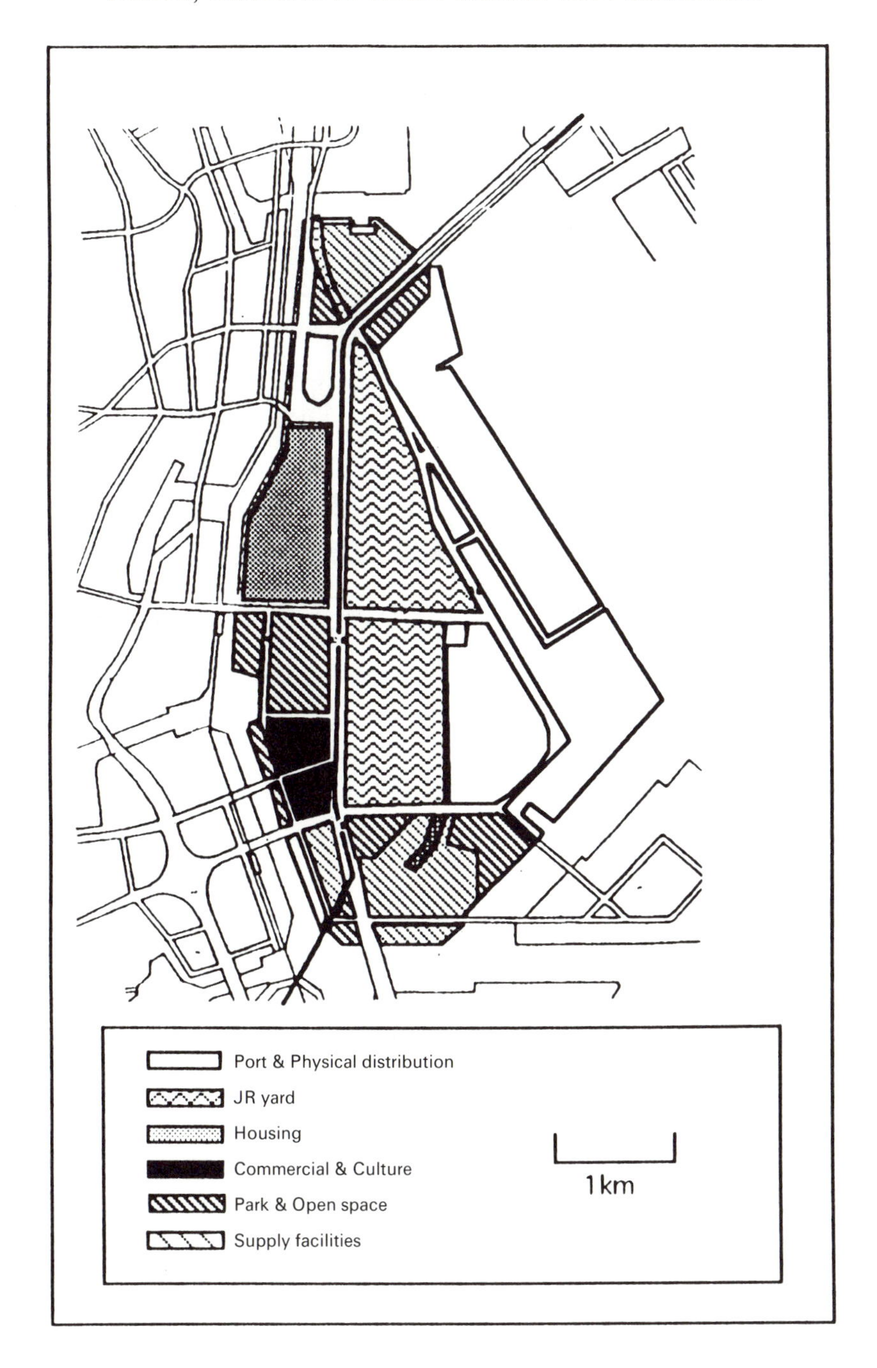

Figure 7.2 Tokyo: land use in Yasio Park Town

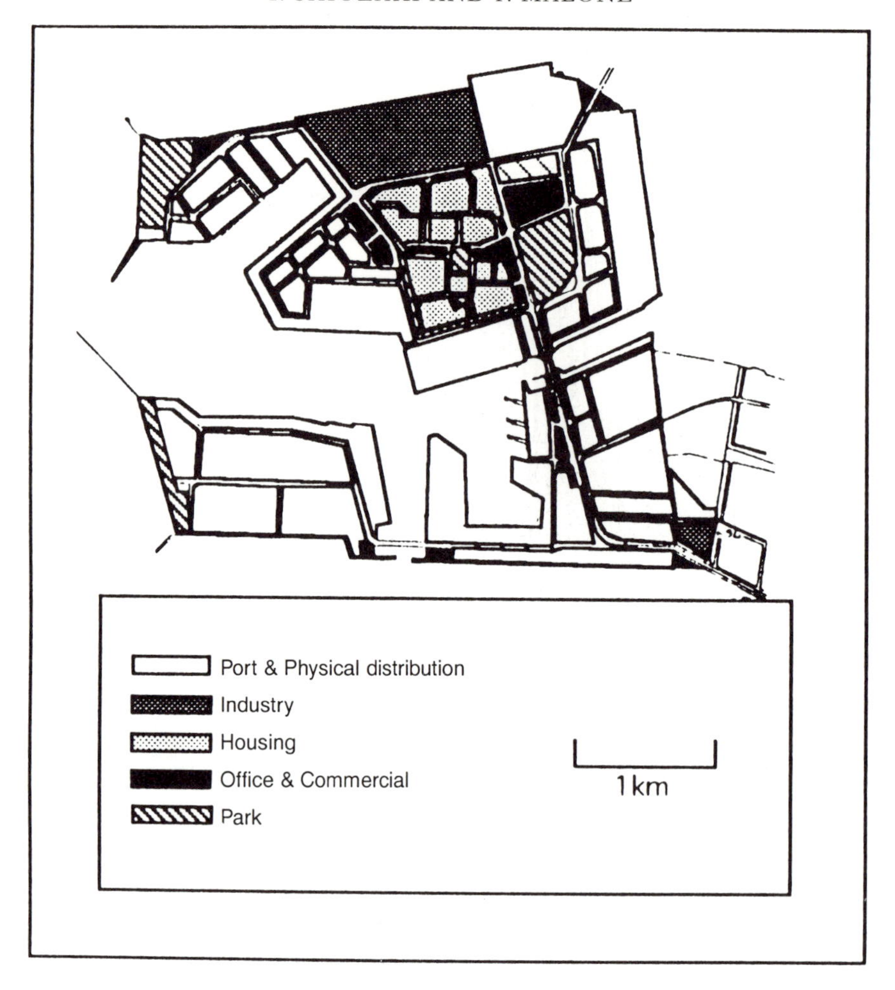

Figure 7.3 Osaka: land use in Nanko Port Town

in all three projects is relatively small, especially in comparison with the areas given over to port, distributive and industrial facilities. While housing covers between 3.8 and 5.2 per cent of the total land area in the three projects, port and quasi-industrial facilities cover between 53.2 and 64.6 per cent. Thus, these are not 'residential islands'. Neither are they housing suburbs of a parent city. They are port facilities, with a high level of allied distributive and industrial functions that accommodate housing, recreational and allied commercial functions. In this respect, the island cities cannot be compared with other, later, marine developments where the primary emphasis is on office space, housing, leisure and, possibly, conference and

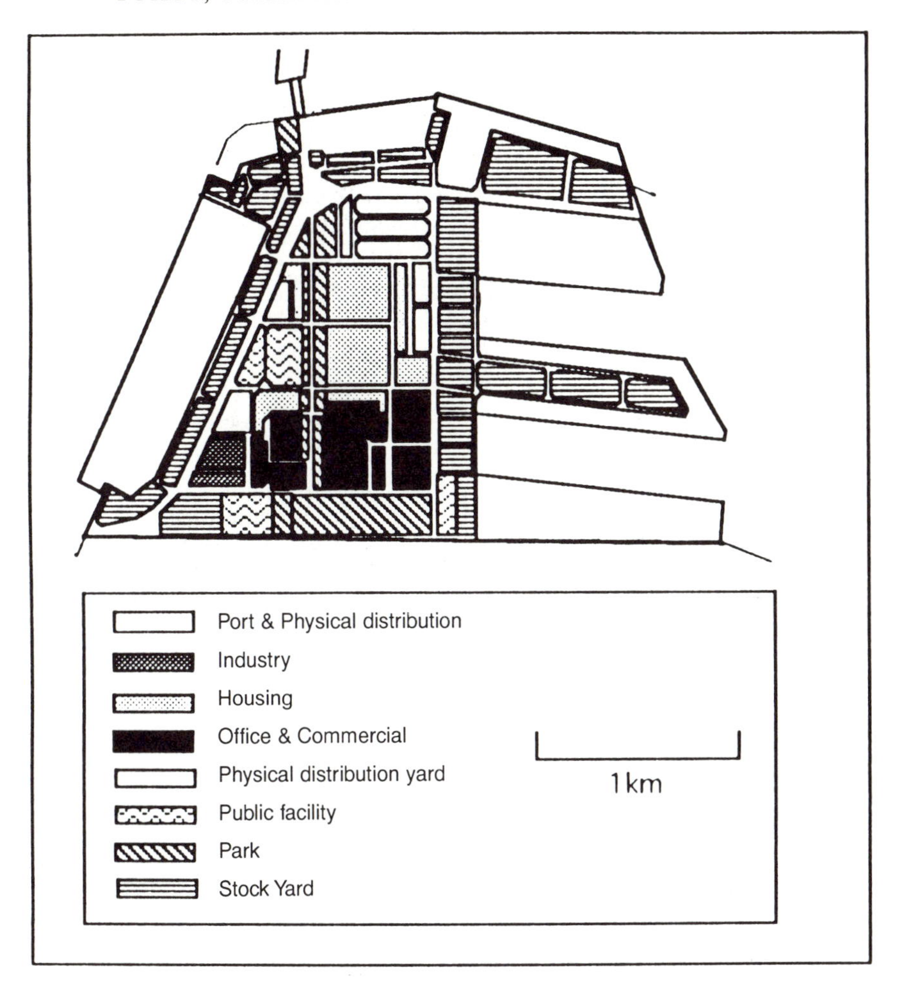

Figure 7.4 Kobe: land use on Port Island

exhibition facilities. Moreover, the nature of these three island cities is such that they may be less deserving of the title of 'city' than other, more recent, proposed Japanese marine projects.

The three projects are also atypical in that they are plugged into the port/industrial sector of the economy. Island cities built around office functions relate to the service sector, and may embody Japanese ambitions to operate within the global economy and the world of international exhibitions and urban marketing (see Chapter eight).

It should also be noted that these island cities are not isolated developments. They are surrounded by other recent waterfront and land-reclamation projects, and may become part of a complex of marine developments. For example, there is a proposal to build another artificial island off the southern end of Port Island in order to accommodate a new airport. Near Nanko Port Town, a new airport (Kansai) has been built on an artificial island with an area of 500 ha. Reclamation is also under way to extend the northern edge of Nanko Port Town to provide convention facilities and offices. These later projects are typical in that they emerged during a boom period for waterfront development – in the middle and late 1980s. They are also typical in that they put further strains on transport systems and the ecology of the areas in which they are located.

Although they vary in terms of their detailed planning, the three island cities share common origins and planning characteristics. They reflect the drive for efficient port and distributive facilities to serve escalating levels of industrial activity and consumption. They are built according to rigorous zoning policies that include sharp divisions between land uses. There is an emphasis on hierarchical road systems and the development of rail and marine transport networks. Zoning and transport policies work together in that major transport elements separate different land uses (Fig. 7.5). Housing is high density and high rise. Residential zones are discrete (not to say isolated) and sharply defined by roads or landscaping. Housing blocks are arranged in patterns that reflect the bureaucratic distribution of devel-

Figure 7.5 Kobe: major avenue on Port Island
Source: Kobe Institute of Urban Affairs (1981)

opment contracts. There is also an emphasis on the provision of open space and path systems. Where they are not 'themed' or crowded with amusements, parks are designed to resemble 'natural' (though obviously contrived) environments. Modernist planning and architecture provide an opportunity to pursue urban-design goals and social values in the open spaces between buildings, through approaches to landscaping and movement and the design of signs and street furniture.

The way in which design goals are pursued through the provision of open space, or through the 'voids' of a master plan, reflects the fact that various developers design and contribute buildings according to a master plan. This development process means that landscaping is necessary to improve the urban environment. On the whole, however, planning ambitions are geared to achieving quantitative rather than qualitative planning goals, and to the introduction of new technologies, such as pneumatic waste disposal.

HOUSING

In each of the three projects the housing element is essentially located within a single zone, although it is important to distinguish between residential zones which incorporate retail and social facilities and the individual housing areas within those zones. In Nanko Port, for example, the residential zone incorporates four housing areas, separated by commercial and social facilities, neighbourhood parks and schools (Figs 7.6 and 7.7). The goal of housing 40,000 people, with all the associated facilities, on 100 ha of land has resulted in high housing densities and mainly fourteen-storey apartment blocks. The residential zone is planned as a modernist 'super-block', although its population is that of a town. It is organized on the basis of quantitative criteria relating to the provision of schools, retail outlets, parks and other elements. There is an emphasis on the functional relationships between the components of the residential zone, and on movement systems that range from the shortest pedestrian routes to the light-rail system that provides the main connection to the mainland.

Typically, housing is built for sale and rent and may be developed by city authorities, the Housing and Urban Development Corporation, private developers or private corporations that provide housing for employees. Variations in rents, tenure and the methods of providing housing are intended to result in differences in social status and to maintain the balance and stability of the population in terms of age structure and the rate at which housing changes hands. As Nanko Port Town demonstrates, the tendency to divide housing areas into parcels of land, which are given over to different types of public and private developers and designated for housing for rent or sale, provides the basis for the spatial layout and the architecture of housing areas.[4]

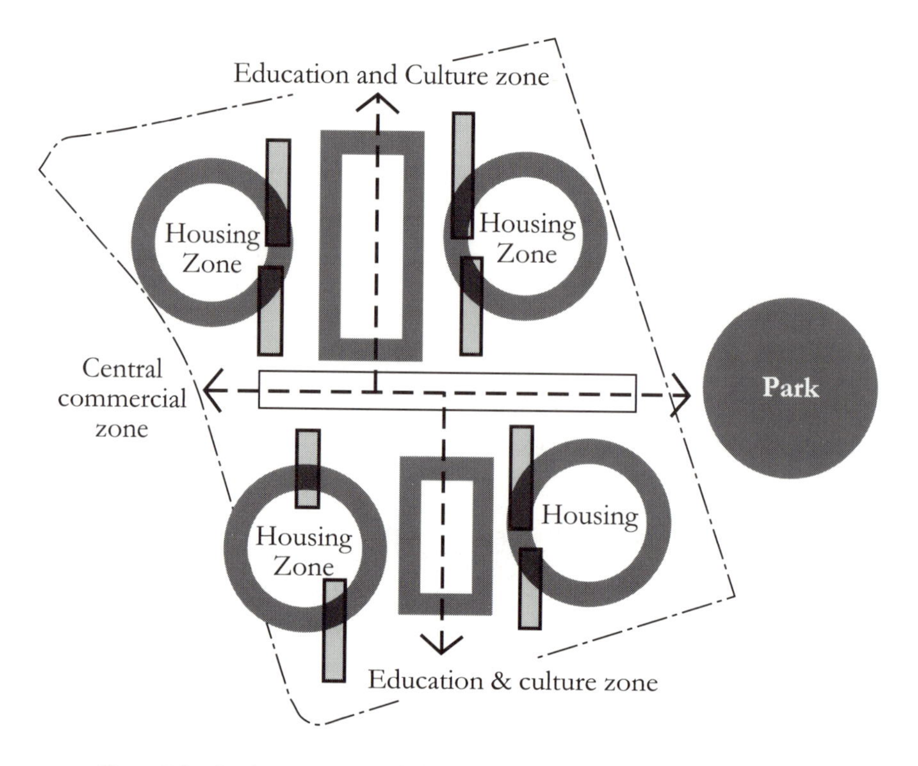

Figure 7.6 Osaka: structure of the residential zone of Nanko Port Town
Source: Osaka Municipal Government (1985)

Because of the size and nature of residential zones in the three island cities, they are dwarfed by the other functions. Mainly, they are isolated by the emphasis given to port facilities and allied transport systems that hug the waterfront and force housing into internal 'walled' zones. In effect, housing areas are doubly isolated in that they are islands within islands. In Nanko Port Town, this is reflected in a master plan that treats the residential zone and its allied facilities as a self-contained 'city' (Ando and Okada 1985: 18–23).

Although the ratio of housing to total land area is relatively constant (roughly 3–5 per cent) housing densities vary between the three projects. Because residential zones vary in size and composition, densities vary greatly according to whether they are assessed on the basis of entire residential zones as opposed to individual housing areas. On the basis of residential zones that include allied facilities, Port Island has a density roughly twice that of Yasio Park Town or Nanko Port Town (where the population density is 400 people/ha). However, the differences in housing

Figure 7.7 Osaka: detailed plan of the residential zone of Nanko Port Town
Source: Osaka Municipal Government (1985)

densities are less extreme if assessed on the basis of housing areas as opposed to entire residential zones. At 295.4 dwellings per hectare (896.8 people/ha), the planned density is again highest for Kobe's Port Island – the smallest of the three projects (Fig. 7.8). Tokyo's Yasio Park Town has a density of 202.6 dwellings per hectare (692.3 people/ha). Nanko Port

Figure 7.8 Kobe: housing on Port Island

Town, the largest of the three islands, and with a planned population that is roughly twice that of the other two projects, has a density of 205.3 dwellings per hectare (821.3 people/ha)(Table 7.2).

The profiles of the inhabitants of the three island cities resemble one another closely. Nuclear families predominate in each (Table 7.3). Householders tend to be in their thirties or forties. Roughly 60 per cent of heads of households work for private companies; a further 11–16 per cent are self-employed. Tokyo's Yasio Park Town stands out in having more than 40 per cent of dual-income families – a much higher proportion than in Nanko Port Town and Port Island. However, the ratio of heads of households employed on their respective islands was only 3.4 per cent in Yasio Park Town, 10.3 per cent in Nanko Port Town and 11.4 per cent in Port Island.

The percentage of inhabitants who were satisfied with their housing conditions was highest in Tokyo's Yasio Park Town (70 per cent). This compares with 47 per cent in Port Island, which has the highest residential densities, and 43 per cent in Nanko Port Town.[5] In that housing densities were relatively low in Yasio Park Town, it could be argued that density is a factor that may influence levels of satisfaction with housing in the three projects.[6]

On balance, the data relating to housing in all of the three projects

Table 7.2 Tokyo, Osaka and Kobe: housing in three island cities

Island city	*Yasio Park Town*	*Nanko Port Town*	*Port Island*
Parent city	*Tokyo*	*Osaka*	*Kobe*
Total area (ha)	679.8	936.8	436.0
Target population	18,000	40,000	20,000
Total area of residential zones (ha)	41.0	100.0	23.0
Housing site area (ha) and ratio to general land use (%)	26.0 (3.8%)	48.7 (5.2%)	22.3 (5.1%)
Planned densities:			
(a) dwellings;	692.3	821.3	896.8
(b) persons per ha of built area	202.6	205.3	295.4

Sources: Tokyo Metropolitan Government, Housing and Urban Development Corporation (1980) *Master Plan of Shinagawa Yasio Housing Estate*; Port and Harbour Bureau City of Osaka (1983) *Osaka Nanko*; Kobe City Development Bureau (1979) *Port Island*
Note: Planned densities are based on housing areas including immediate open space, pathways, etc., but excluding trunk roads.

indicate high levels of dissatisfaction. Even in Tokyo's Yasio Park Town, 35 per cent of households wanted to move out. Approximately 40 per cent of households in Nanko Port Town and 49 per cent in Port Island expressed an urge to move. In short, between 35–50 per cent of residents were clearly dissatisfied with the housing in the three projects.

The extent of residents' dissatisfaction with their housing may indicate problems with the nature or quality of dwellings and residential zones. However, it could also reflect dissatisfaction with the quality of social and other facilities. Thus, when conducting research into the living conditions in the three island cities, transportation, shopping and medical facilities were taken into account. Together with housing, these were thought to have the greatest influence on the overall levels of satisfaction for residents.

TRANSPORT

The efficient distribution of goods underpins the economic role of island city projects that are based on the provision of advanced port facilities. Efficiency is dependent on hierarchical systems of transport, which sift different types of traffic, and on the location of port facilities, wharfs and container terminals in relation to transport systems. For example, large

Table 7.3 Tokyo, Osaka and Kobe: households in three island cities

Island city	*Yasio Park Town*	*Nanko Port Town*	*Port Island*
Parent city	*Tokyo*	*Osaka*	*Kobe*
Total area of islands (ha)	679.8	936.8	436.0
Target population	18,000	40,000	20,000
Target no. of households	5,270	10,000	6,588
Actual population (1985)	16,899	28,600	14,000
Composition of households			
Families with children (%)	75.7	79.8	58.3
Couples without children (%)	12.1	13.8	24.6
Single-person households (%)	5.8	4.6	11.1
Other households (%)	6.3	1.7	6.1

Sources: Tokyo Metropolitan Government, Housing and Urban Development Corporation (1980) *Master Plan of Shinagawa Yasio Housing Estate*; Port and Harbour Bureau City of Osaka (1983) *Osaka Nanko*; Kobe City Development Bureau (1979) *Port Island*
Note: Data on household composition are based on data gathered at Port Island (Kobe) in 1984, Nanko Port Town (Osaka) in 1985 and at Yasio Park Town (Tokyo) in 1989.

container terminals are strategically located in terms of main roads and access to the mainland.

The land-use structure of the islands is based on the overwhelming importance of port functions and their relationship to regional transport systems. However, attention is paid to the movement of pedestrians and bicycles in residential zones. In Nanko Port Town, for example, cars are excluded from housing areas in order to reduce noise, accidents and pollution. This system allows people to walk between housing, schools and other facilities. However, some residents commute to peripheral car-parking areas by bicycle or motorcycle, and traffic restrictions can frustrate residents' access to cars in medical and other emergencies.

Whatever the quality of the transport systems within the residential zone in Nanko Port Town, the overall plan of the island is typical, in that the residential zone is isolated within the overall plan, while the cost and convenience of transport systems connecting the island to its parent cities are important factors in determining the quality of life on the island.

Japanese new towns built on the mainland are commonly 'dormitory' communities that may suffer, initially, from poor transport links with their parent city and further afield. In economic and physical terms, however, it may be easier to create adequate transport links to inland towns. Moreover, psychologically, island cities impose a sense of restricted movement, which is exacerbated by the fact that transport links to the mainland are prone to congestion, minor disruptions, serious accidents and potential disasters.

The need for good transport links between island cities and parent cities is indicated by the number of people who may commute to work on the mainland (Table 7.4). Of the three islands, Port Island is closest to its parent city (Table 7.5). However, only one bridge connects Port Island to Kobe. This 200 m bridge is 50 m above sea-level and is difficult to cross on foot or by bicycle. Thus, movement between Port Island and Kobe is limited to motor traffic and one public-transport link – a light-rail connection known as the Portliner. Nanko Port Town is also connected to Osaka by a surface light-rail link (the New Tram). Nanko Port Town's rail link is operated by a public body. In Port Island the system is operated by a semi-public body. As such, it is not awarded any special operating subsidies.

Port Island residents, some of whom use the Portliner rail system on a daily basis, complain of high fares and the lack of other transportation options. According to residents, the Portliner is unsatisfactory in terms of both the system and its management. Essentially, it is a medium-sized, low-speed rail system running in a one-way loop (Fig. 7.9). Although it looks modern, it is slow, and some passengers have to detour around the loop. It also suffers from congestion and periodic overcrowding because it serves Port Island's amusement park, stadium, business facilities and a convention hall, in addition to the port and its allied facilities.

Port Island's problems result from an attempt to service the widest possible area at the smallest possible cost. There is an obvious need to augment the island's public transport services. Of the three island cities, only Port Island has no bus service. However, Kobe's city government does not want

Table 7.4 Tokyo, Osaka and Kobe: commuting to parent cities

Island city	*Yasio Park Town*	*Nanko Port Town*	*Port Island*
Parent city	*Tokyo*	*Osaka*	*Kobe*
Total area (ha)	679.8	936.8	436.0
Target population	18,000	40,000	20,000
Percentage of heads of households employed on their home island			
	3.4	10.3	11.4
Percentage of commuters working on the mainland			
	96.6	89.7	88.6

Sources: Tokyo Metropolitan Government; Port and Harbour Bureau City of Osaka; Kobe City Development Bureau
Note: Data were collected at Port Island (Kobe) in 1984, Nanko Port Town (Osaka) in 1985 and at Yasio Park Town (Tokyo) in 1989.

Table 7.5 Tokyo, Osaka and Kobe: linking island and parent cities

Island city	*Yasio Park Town*	*Nanko Port Town*	*Port Island*
Parent city	*Tokyo*	*Osaka*	*Kobe*
Total area (ha)	679.8	936.8	436.0
Target population	18,000	40,000	20,000
Distance/time between island/parent city centres			
	9.5 km/30 min.	10.0 km/40 min.	2.5 km/10 min.
No. of bridges	6	2	1
Main bridge	120 m	500 m	200 m
Rail access	Monorail	New Tram	Portliner

Sources: Tokyo Metropolitan Government; Port and Harbour Bureau City of Osaka; Kobe City Development Bureau
Notes: Commuting times are approximate. The data given for the bridges indicate approximate distance between shorelines for the main bridge that is nearest the mainland. Yasio Port Town has additional connections to other islands, and a tunnel.

to destroy the monopoly enjoyed by the Portliner by introducing a competitive bus service.

One of the issues raised by respondents (in 1984) with regard to transport links to the mainland was the fear of being stranded in an emergency. These fears proved to be well founded when Kobe suffered an earthquake in the early hours of 17 January 1995. The bridge to the mainland was slightly damaged and access was restricted for a period of days. The rail system, however, was severely damaged and was out of use for seven months. During that period, the island's roads and the bridge to the mainland had to absorb additional traffic due to the closure of the rail system and traffic generated by the disaster and its aftermath. Port facilities on the island were badly damaged. Some of the island's high-rise housing blocks, although designed to counteract the effects of earthquakes, were cracked and tilted as the ground buckled, and the island's soil suffered liquefaction effects. Thus, the damage sustained by the Portliner (and the light-rail Rokkoliner that serves Kobe's second island city) created a sense of isolation and added to the psychological stress and hardships suffered by residents. Moreover, because Kobe's main hospital had been relocated to Port Island, the damage to the island's transport links severely reduced access to the main hospital for the residents of the mainland.

In Nanko Port Town, two bridges provide access to Osaka and the mainland. However, commuters regard road connections to Osaka as inadequate and expensive. The shortest connection is via an expensive toll bridge. Thus, the principal connection to Osaka is the New Tram surface

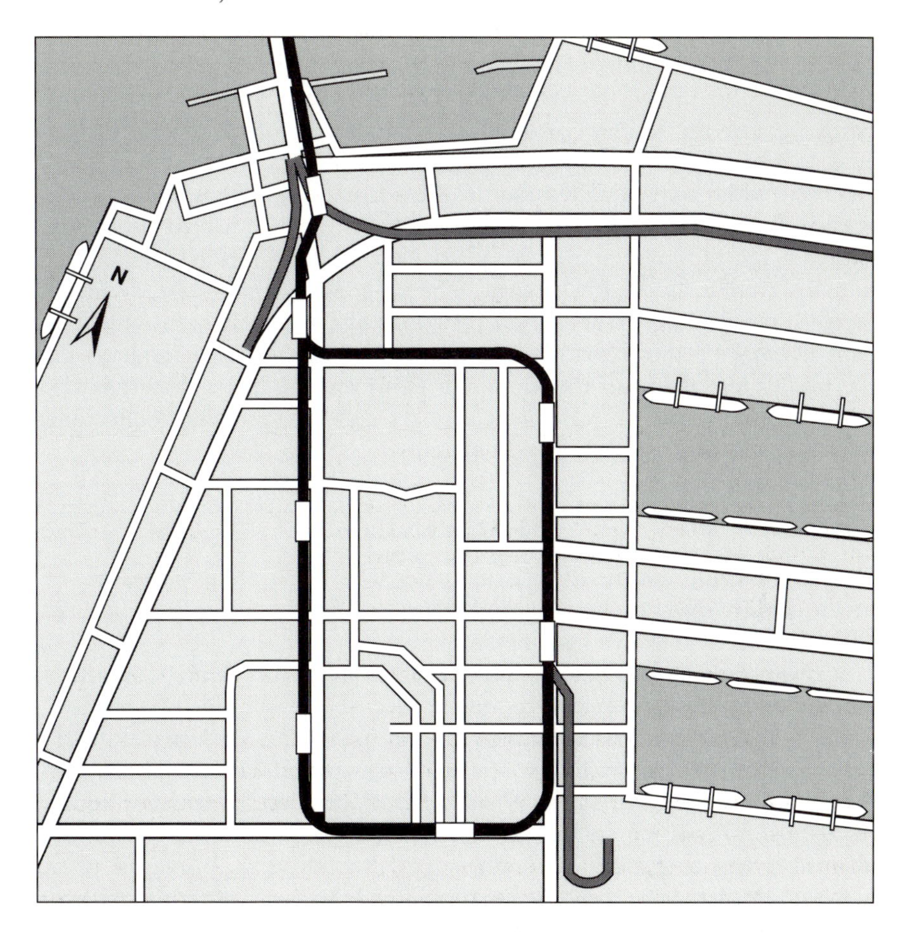

Figure 7.9 Kobe: housing and rail system on Port Island
Source: Kobe New Tram Company

rail system. There is a bus service but this is infrequent. Both the New Tram system and the buses are operated by Osaka's city government.

The New Tram runs through the centre of Nanko Port Town, and while many of the island's residents use it frequently, some complain that it takes more than forty minutes to undertake the 10 km trip to the terminus in Osaka. Commuters also complain of inconvenient transfers between transport systems in Osaka. Although Nanko Port Town has two bridges, the New Tram crosses to Osaka by means of the bridge which is furthest from the city centre, and terminates in Osaka at a subway station. Because of the type of track it uses, the New Tram cannot run on Osaka's subway tracks. Transferring between systems in Osaka requires passengers to move, via escalators, between the elevated New Tram terminal system and the subway station two floors below ground level.

There are six bridges between Yasio Park Town and the mainland. This is partly because the Keihin Canal, which separates the island from the mainland, is relatively narrow (roughly 120 m). In addition to the bridges, there are two public transport systems in the form of a monorail and a bus service. Both of these take about thirty minutes to reach Tokyo city centre, although the monorail station in Yasio Park Town is located roughly ten minutes' walk from the town centre. The monorail system is also disadvantageous in terms of the location of stations and relatively high fares. Perhaps because of these disadvantages, more Yasio Park Town residents use the bus system than use the monorail. However, the bus service has also attracted complaints of infrequent services and inconvenient connections to other parts of Tokyo's transport system.

SHOPPING

Shopping facilities on the three island cities were evaluated as a measure of urban quality. Like the housing and transport facilities, shopping facilities elicited relatively negative responses.

It might be assumed that the retail facilities of marine cities would provide food and the convenience goods to meet the daily needs of shoppers, rather than, for example, expensive clothes, furniture, or electrical goods, for which people are willing to travel to the parent city. In Japan, it is particularly important that shops selling food and basic necessities are located close to housing areas. The Japanese lifestyle is such that people are inclined to buy food on a daily rather than a weekly basis. In marine cities, however, retailers have a natural monopoly, and the fact that there is little or no competition affects prices and levels of service. In Port Island, for example, there are only two large supermarkets, an affiliated smaller supermarket and a convenience store, to serve a planned resident population of 20,000. Thus, the pattern of retailing is not well geared to the sale of basic goods for daily use and the island has only a few small local shops and specialized stores (Table 7.6).[7]

Most of the stores on Port Island are managed by the Daiei Corporation, Japan's biggest retail company. Retailing in Port Island is dominated by Daiei to the point that the island is sometimes referred to as 'Daiei Land'. The Daiei Corporation pioneered the American-style supermarket in Japan and, initially, it offered discount prices. Today, however, its retailing strategy varies with the nature of local market conditions and the opportunities for profit. Since it has no competition in Port Island, Daiei may sell goods in the island's supermarkets at prices that are higher than in Daiei supermarkets on the mainland.[8]

The shopping situation in Nanko Port Town is comparable with that of

Table 7.6 Tokyo, Osaka and Kobe: shopping in three island cities

Island city	*Yasio Park Town*	*Nanko Port Town*	*Port Island*
Parent city	*Tokyo*	*Osaka*	*Kobe*
Total area (ha)	679.8	936.8	436.0
Target population	18,000	40,000	20,000
Supermarket	1	1	3
Convenience store	—	1	1
Other store	—	16	Some
Street vendor	Some	—	Few

Sources: Tokyo Metropolitan Government; Port and Harbour Bureau City of Osaka; Kobe City Development Bureau
Notes: Data gathered at Port Island (Kobe) in 1984, Nanko Port Town (Osaka) in 1985 and at Yasio Park Town (Tokyo) in 1989

Port Island. There is one retail complex in the centre of Nanko Port Town, which contains a supermarket and some smaller stores. There are four additional stores (for rice, medicine, liquor and basic foods) within the island's residential zone. Thus, there are more retail outlets than on Port Island, but the density of retail space is still relatively low, i.e. 2.4 stores per 1,000 residents. This is much less than in the parent cities of Kobe, or in Osaka or Tokyo where the ratios of retail stores per 1,000 of the residential population are 14.1, 19.5 and 12.8 respectively.

In Tokyo's Yasio Park Town there is one large shopping mall and a few street vendors. The main store in the shopping mall is a supermarket managed by Daiei. Respondents in Yasio Park Town were more negative about shopping facilities than in Port Island and Nanko Port Town. This contrasts with levels of satisfaction with housing, which were highest in Yasio Park Town.

In general terms, roughly 90 per cent of the respondents in each of the three island cities expressed dissatisfaction with shopping facilities. Over 70 per cent complained about the limited range of shops and goods available, and over 60 per cent that prices were too high. They also complained about early closing times. This particularly affected commuters.

Solutions to the problems associated with retailing may vary. In Port Island and Yasio Park Town street vendors already provide some improvement in shopping facilities. This is particularly evident in Yasio Park Town where fresh foods and 'fast food' are sold on the street. A high percentage of island residents (for example, 68 per cent in Yasio Park Town) solve the problem by shopping on the mainland. However, this is not an ideal solution, even for those who work on the mainland, and it is a poor substitute for better facilities in the three marine cities.

MEDICAL FACILITIES

The range and availability of medical facilities on the three island cities were also examined (Table7.7). These are significant factors in the quality of life in marine cities, where the range of medical facilities and the ease of access to those facilities are tested by everyday and emergency needs.

The extent to which a marine city provides a self-contained system of medical support may be assessed by the level of everyday access to a wide range of specific facilities. In an emergency, however, access to medical facilities may be a question of life or death, and doubts about emergency facilities can provoke deep anxiety among the residents of island cities.

In general, residents considered that the medical facilities on all three artificial islands were inadequate. Nanko Port Town and Yasio Park Town are typical of many marine cities, in that they have neither a hospital nor an emergency clinic. They are also typical in that they have a relatively low number of medical practitioners and a shortage of specialist help, for example for ophthalmic treatment. In this respect, Port Island is exceptional in that it has a good hospital, which is centrally placed and is managed by the local government. But here the exception ends, because apart from this hospital there are relatively few medical practitioners.

The Central Municipal Hospital in Port Island is well equipped. However, it is a regional hospital. As stated above, it was built to replace the old Central Municipal Hospital in Kobe. Essentially, the regional hospital facilities were moved from the city centre to Port Island and were equipped with helicopter access for emergency cases. The move was made against the wishes of the citizens of Kobe and is said to have involved political wrangling by 'insiders'. The hospital is part of a larger complex, which includes a shopping centre and a hotel managed by Daiei.[9]

The relocation of the hospital was opposed on the basis that it would

Table 7.7 Tokyo, Osaka and Kobe: medical facilities in three island cities

Island city	*Yasio Park Town*	*Nanko Port Town*	*Port Island*
Parent city	*Tokyo*	*Osaka*	*Kobe*
Total area (ha)	679.8	936.8	436.0
Target population	18,000	40,000	20,000
Number of medical facilities and general hospitals			
	10	14	8
	—	—	1

Sources: Tokyo Metropolitan Government; Port and Harbour Bureau City of Osaka; Kobe City Development Bureau
Notes: Data gathered at Port Island (Kobe) in 1984, Nanko Port Town (Osaka) in 1985 and at Yasio Park Town (Tokyo) in 1989

cause inconvenience to mainland patients and visitors, and would increase pressure on the island's only rail link. In this respect, it is interesting that some families have moved to Port Island because of their dependence on the hospital. The relocation of the hospital was also opposed on the grounds that Port Island may be vulnerable to earthquakes and is serviced by only one bridge. These fears were realized when the earthquake reduced access to the hospital in 1995. In the period immediately after the disaster, the damage sustained by the bridge and rail links to the mainland meant that the hospital was inaccessible except by helicopter. The damage suffered by the rail link frustrated access to the hospital over a period of seven months.

The problems experienced with the medical facilities in all three marine cities, whether in terms of a shortage of doctors or access to emergency or general treatment, have caused widespread inconvenience and concern. The ratio of medical practitioners was found to be 1:2,000 people in Nanko Port Town. This compares with 1:860 in Osaka, the island's parent city.

ENVIRONMENT AND AMENITY VALUES

As new environments, island cities should represent an opportunity for good planning and the creation of recreational facilities and public amenities. The least that might be expected from an island city is that its population might enjoy the waterfront and its recreational facilities. Unless a marine development can provide its residents with good waterfront amenities, its unique potential is wasted.

Of the three island cities, only Tokyo's Yasio Park Town possesses adequate public open space and waterfront amenities. Here the residential zone adjoins both a park and the Keihin Canal, and there are two other major areas of parkland and open space. The island's parks accommodate a range of activities, and the population has access to some 4 km of shoreline (Table 7.8).

In Nanko Port Town the situation is less successful. Although there are 2 km of accessible waterfront, this is flanked mainly by port and distributive functions, and is located too far from housing areas to allow casual access to the waterfront or good views between housing and the sea. On the plus side, there is a bird sanctuary, a fishing deck and an artificial bathing beach; although these are located away from the residential zone and are obviously artificial. The bird sanctuary is the most successful of Nanko Port Town's waterfront projects. It is a huge marsh (19 ha) made up of a good balance of reclaimed land and sea-water. It contains two observatories. The island's fishing deck is built along a bank which encloses a sandy bathing beach decorated with some tropical huts made from palm trees. The sand making up the beach is imported and both the sand and adjoining pool are cleaned

Table 7.8 Tokyo, Osaka and Kobe: recreational facilities in three island cities

Island city	*Yasio Park Town*	*Nanko Port Town*	*Port Island*
Parent city	*Tokyo*	*Osaka*	*Kobe*
Total area (ha)	679.8	936.8	436.0
Target population	18,000	40,000	20,000
Total area of parks (ha) (% of total area)			
	92.0 (13.5)	81.1 (8.7)	24.0 (5.5)
Access to public waterfront (km)			
	4.0	2.0	0.5
Parks and recreational facilities			
	Obikaihin Park Bird sanctuary	Fishing park Bird sanctuary	Three parks Portopia Land bathing beach

Sources: Tokyo Metropolitan Government; Port and Harbour Bureau City of Osaka; Kobe City Development Bureau

by machinery. This is an artificial tropical paradise maintained by machinery and overlooked by a power station (Fig. 7.10).

On Port Island only 500 m (or 3.5 per cent) of the island's waterfront is accessible to the public. This is located in a small park under the bridge which links the island to the mainland. The remainder of the island's 14 km of waterfront is taken up mainly by port facilities. Moreover, as in Nanko Port Town, there is an unsatisfactory relationship between the residential zone and the sea.

In each of the three island cities, parks or green areas comprise between 5.5 and 13.5 per cent of total area. These ratios are not high relative to Japanese planning standards. In inland new towns the proportion of parks and peripheral green areas is typically 20–30 per cent. Moreover, there are problems with the distribution of green areas in the island cities. For example, there are three parks in Port Island, which are located some distance from the island's housing and which cannot be reached without crossing busy roads. Adjoining the biggest of the three parks, and a long distance from any housing, is an amusement park called Portopia Land. This takes its name from the *Portopia Exhibition*, which was staged by the Kobe city government in 1981 to launch the Port Island project. The site of the amusement park, which formed part of the exhibition, was orientated towards the local commercial area rather than the island's housing area.

Overall, the parks at Nanko Port Town and Yasio Park Town have turned out well. In Nanko Port Town, the residential zone is surrounded by a green 'embankment' which isolates it from the port and industrial facilities

Figure 7.10 Osaka: artificial beach at Nanko Port Town

and forms a path leading to a large park.[10] When asked about this layout, 95 per cent of the respondents agreed that there is 'a lot of green', but this was qualified by negative reactions to the quality of park and landscaped areas. A major cause of dissatisfaction is an artificial stream, operated by pumps. Thus, perhaps one lesson to be learned from Nanko Port Town is that imitations of nature in the form of artificial bathing beaches and mechanical streams may be rejected by a 'captive' population.

Japan's island cities also demonstrate that the planning of new mixed-use developments raises problems that relate both to the environment and to issues in urban design. For example, there is a need for careful planning and management where public open space and housing areas adjoin industrial and port facilities. Housing may abut open yards with large stocks of dangerous chemicals. Domestic traffic must contend with heavy-goods traffic serving port and allied facilities. In terms of the combination of different uses and the management of traffic, all of the three marine cities examined showed signs of bad or insensitive planning. This is evidenced, for example, in the heavy traffic on the ring roads, which are difficult to cross; in the movement of people between housing and amenity areas; and, as in Yasio Park Town, in the close proximity of housing and major roads.

IMAGE AND REALITY

Each of the three island cities was marketed as a 'new frontier' and a 'future paradise'. However, the experiences of those who live in these marine cities suggest that there is a wide gap between reality and the marketing imagery used to promote these projects. In terms of the quality of their planning and the facilities that they offer, the projects do not provide the ideal living environments promised in advertisements. It could be argued that these island cities need more time to mature. However, research leaves little room to hope that they will become more satisfactory over time. In fact a second round of surveys in Port Island and Nanko Port Town (conducted in 1990 and 1992, or six and seven years after the first surveys) showed an increase in negative responses to questions about transport, shopping, medical facilities and factors which determine the quality of life on these islands.[11] To date, there has been little emphasis on improvement. Some new parking facilities have been constructed on Port Island and in Yasio Park Town to rectify earlier errors in the estimation of parking needs. But it is perhaps indicative that the number of cars in the three cities has actually increased as residents have sought to overcome transport and mobility problems. New parking facilities cut into green areas, which diminishes further the amenity value of public open space. Meanwhile, as trunk roads are expanded to carry increasing levels of traffic in port areas, this additional traffic brings more air pollution, noise and accidents. New road construction may ease traffic congestion temporarily, but the extensive road programmes that feed Japan's passion for large marine developments will escalate environmental and other costs.

THE NEED FOR ISLAND CITIES

Marine developments on new or reclaimed land are a significant feature of urban development in Japan's waterfront cities.[12] Between 1960 and 1985, a total of 7,300 ha of land was reclaimed in Osaka Bay, and a remarkable 18,800 ha was reclaimed in the Tokyo Bay area.

It is not possible to explain the tendency to reclaim land simply in terms of a shortage of redundant or development land. In the Osaka Bay area, for example, there was (in 1990) a total of 1,263 ha of redundant and under-used industrial land in need of redevelopment (Foundation of Osaka Science and Technology 1988). However, a total of 4,041 ha of land was earmarked for reclamation and the bulk of this new land (3,405 ha) was in the form of artificial islands in Osaka Bay. The fact that the area earmarked for reclamation was roughly three times the area of land already available for redevelopment may testify to the pressure on development land in Osaka, but the question remains as to why more of that pressure is not directed towards existing land.

One of the reasons why the current approach to marine developments has found favour with metropolitan governments is that it can be presented as a potential solution to urban problems, notably problems related to congestion and urban expansion. Another reason is that recent large marine developments represent a welcome shift away from the pattern of waterfront reclamation and port development that began after the World War II. In the period up to the early 1970s, there was a widespread tendency for port, distributive and industrial functions to replace shoreline recreational areas and fishing ports. This process increasingly deprived the public of recreational facilities and access to the waterfront. It also left a legacy of serious air and water pollution. In the 1970s, following a period of vociferous opposition to various forms of pollution in Japan, there was a reaction against the location of heavy industry within urban areas. While the nature of industrial production has altered in the period since the early 1970s, planning authorities have sought to restructure land-use planning to reflect environmental concerns. The planning of Port Island near Kobe is a good example. Initially, the project was intended to accommodate port facilities and heavy industries. However, the Port Island project was adapted to provide mixed-use development and to accommodate elements such as a hotel, convention centre, stadium, open space and public amenities. Although port facilities continue to constitute a major part of many marine developments, the tendency towards mixed-use projects is seen as a welcome step away from homogeneous port/industrial projects.[13]

It would, however, be naïve to suggest that mixed-use marine developments follow from environmental concerns when these developments are accused of damaging the natural environments of parent cities. In Tokyo, for example, it is argued that proposed marine developments will further exacerbate problems associated with the city's overloaded transport networks, its stretched housing stock and its environment (Chapter eight). In addition, leaving aside their relationship with the parent cities, it is clear that mixed-use marine projects may pose environmental problems associated with, for example, air pollution, or the proximity of housing to stores of dangerous chemicals or materials. Arguably, however, the most pressing environmental concerns spring from the vulnerability of large marine developments to earthquakes and the fact that they may compound the human and physical costs of earthquakes in Japanese cities.[14]

While reactions to older monofunctional port/industrial developments may play some part in the promotion of marine developments, clearly profit is the major factor behind the involvement of private developers. Metropolitan governments, on the other hand, may be driven by visions of an expanding urban economic base, enlarged commercial cores, increased housing stocks or the provision of recreational or other facilities. The temptation to use new and public land to expand the physical and economic structures of cities is enhanced by the fact that metropolitan governments are caught up in a race for urban economic status. Metropolitan governments are also under pressure to find solutions to Japan's urban problems.

In summary, it may be concluded that artificial islands and large marine developments do satisfy the need for new port facilities, new technologies and larger ships – factors which are important in a country dependent on foreign trade. In addition, marine projects are driven by the need for new space for commercial functions, housing, recreational and other facilities. They play a major role in the urban economy and contribute directly to economic growth in specific sectors of the economy, notably the construction sector. There are also other factors involved. For example, marine developments satisfy the need for sites for disposing of waste generated by industry and construction – although the need for artificial islands for waste disposal may be exaggerated, if only because tougher regulations could reduce levels of waste.

Given the weight of evidence against them, it is important that the arguments in favour of large marine developments in Japanese cities are not presented as being watertight. The argument that congested cities need marine developments on new land begs questions about the capacity of existing land, which may be in need of redevelopment, to absorb a greater share of development pressures in Japan's waterfront cities. While the need to provide new housing is also given as one of the major reasons for building marine cities, it could be argued that other options are available to solve Japanese housing problems, for example in terms of agricultural or redundant industrial land. Moreover, where the resident population in a marine city is less than the working population, it is difficult to see how a marine city can have anything other than a negative impact on the overall housing stock and transport systems of the parent city. It could also be argued that marine projects are not an ideal or effective solution to housing problems, in that the housing constructed within marine developments can create specific difficulties relating to, for example, access to and from the mainland and to shortfalls in social and welfare facilities. These problems may be particularly acute in the early stages of development, but some persist and are difficult to resolve because of the isolating effect of the sea and planning, which places housing in a secondary and isolated position in the land-use structure.

CONCLUSIONS

In summary, it can be said that Japan's artificial island cities may have negative impacts on surrounding marine and urban environments. They present new dangers in areas prone to earthquakes. Moreover, they may exacerbate rather than relieve the planning problems of the parent city. At the same time, they may fail to provide good urban and housing environments in terms of social and human needs.

All of these factors may be doubly regrettable where large mixed-use marine developments are not necessary to satisfy the demands of urban growth. Leaving aside the need for new port facilities, the most powerful motives for the creation of artificial islands may be the ambitions of urban governments and the fact that marine developments generate large profits for development, construction and other interests. Increasingly, marine projects allow the development industry to invade large areas of new land with the backing of proactive local authorities. Artificial islands and the elevated freeways, bridges and tunnels that link them to the mainland generate large opportunities for construction capital. Moreover, marine developments create space which is absorbed by evolving urban economies. In short, marine cities are big business. They are important to Japan's waterfront cities and its national economy.

Leaving aside their economic role, however, it is open to question whether Japanese island cities offer the best solution to existing urban problems related, for example, to congested central business districts, housing shortages or a lack of open space. Japan's new marine developments are more likely to exacerbate the transport and environmental problems of parent cities. However, the negative aspects and potential hazards of marine cities may be overshadowed by local ambitions for economic growth, particularly where expansion is seen as the key to survival in national or global urban hierarchies.

Given that doubts can be raised with respect to island cities and large marine projects generally, the question arises of why there is such an emphasis on this form of development in Japan. In addressing this question it would be unwise to discount the aspirations of metropolitan governments and private capital, and the power that can be exercised by coalitions of public and private interests.

NOTES

1. Cities such as Tokyo, Nagoya, Osaka, Kobe, Hiroshima, Fukuoka and others have significant reclamation programmes.
2. Moreover, inland new towns may originate from Japan's new town and housing legislation, whereas marine cities may be based on legislation concerned with reclamation (Public Water Surface Reclamation Act).
3. The privatization of development processes in Japan in the late 1980s brought changes in terms of the role of development and property capital.
4. Proposals relating to Nanko Port Town show that the residential area is divided between different types of developers: Osaka Municipal (rent); Housing and Urban Development Corporation (rent and sale); Osaka Municipal Housing Supply Public Corporation (sale); and private corporations (rent and sale) (Ando and Okada 1985: 21).
5. The research for this chapter was undertaken at Port Island (Kobe) in 1984 and 1990, Nanko Port Town (Osaka) in 1985 and 1992 and at Yasio Park Town

(Tokyo) in 1989. Research involved surveys of samples of the populations of the three island cities. Different respondents were used for each stage of the research.

6. There is no direct relationship between housing densities and the occupations or social status of the residents.
7. The specialized stores sell, for example, sports goods, leather goods, clothes, etc.
8. A women's group has compared prices at mainland and Port Island stores.
9. It is not possible, however, to provide evidence that Daiei used its influence to secure the removal of the hospital to Port Island. The original site of the hospital has since been enhanced by the building of a new tunnel and subway station and is now used for a hotel complex, managed by the Daiei Corporation. However, it is clear that moving the hospital to Port Island also benefited the interests that operate the Portliner.
10. In Port Island a similar barrier is used to isolate housing from traffic. In Rokko Island, Kobe's other artificial island city, a barrier (City Hill) is used to isolate housing from industrial facilities.
11. Surveys were undertaken of housing on Port Island in 1984 and 1990 and at Nanko Port in 1985 and 1992.
12. Here the difference between Japan and other countries lies not in the emphasis on waterfront development but rather in the extent of development on reclaimed land. However, the trend towards artificial island projects is now spreading outside Japan. For example, it is intended to create an artificial island, similar to that near Kobe, in the area of Pusan, Korea. This island will be roughly 600 ha in area.
13. The Environmental Conservation Act of the Seto Island Sea was established in 1973. This prohibits reclamation in Osaka Bay unless it contributes to environmental improvements. However, since the act was passed, there has been a lot of reclamation in this area.
14. Generally, all areas of Japan, including the three cities covered here, are vulnerable to earthquakes. The Tokyo metropolitan government, for example, holds that there is a high probability that 40 per cent of newly reclaimed land in Tokyo could suffer 'liquefaction effects' in the event of an earthquake.

BIBLIOGRAPHY

Ando, S. and Okada, T. (1985) *Development of the Nanko Port Town*, Osaka: Osaka Municipal Government.

Foundation of Osaka Science and Technology (1988) *Land-use in Osaka Bay Area*, Osaka: Osaka Centre for Science and Technology (in Japanese).

Housing and Urban Development Corporation (1980) *Master Plan of Shinagawa Yasio Housing Estate*, Tokyo: Housing and Urban Development Corporation.

[Japanese] Ministry of Transport (1986) *Artificial Islands*, Tokyo: Ministry of Transport (in Japanese).

Kobe City Development Bureau (1979) *Port Island*, Kobe: Kobe City Development Bureau.

Kobe City Government (1991) 'Kobe: Port Island, Rokko Island', in R. Bruttomesso (ed.) *Waterfront: Una Nuova Frontiera Urbana*, Venice: Centro Internazionale Città d'Acqua.

Kobe Institute of Urban Problems (1981) *Towards a Marine Cultural City for Kobe*, Kobe: Kobe Institute of Urban Problems (in Japanese).

—— (1986) *Theory and Practice of Island City*, Tokyo: Keiso Shobo (in Japanese).

Port and Harbour Bureau City of Osaka (1993) *Osaka Nanko*, Osaka: Osaka City Government.

Port of Kobe Authority (1987) *Port of Kobe in the Future*, Kobe: Port of Kobe Authority.

Shiozaki, Y. *et al.* (1985) *A Study on Residential Environment of the Port Island in Kobe*, Kobe: Kobe University.

Shiozaki, Y. *et al.* (1986) 'Evaluation of the living facilities on Marine City', paper of the Annual Conference of the City Planning Institute of Japan (in Japanese).

Shiozaki, Y. *et al.* (1990) *A Study on Residential Environment of Marine Cities*, Kobe: Faculty of Engineering, Kobe University.

Tokyo Metropolitan Government (1980) *Master Plan of Shinagawa Yasio Housing Estate*, Tokyo: Tokyo Metropolitan Government.

Tomita, K. (1986) *Changing Osaka*, Tokyo: Tokyo Horeishuppan (in Japanese).

Urban Environment Research Group (1987) *Urban Waterfront*, Toshibunkasha: UERG (in Japanese).

—— (1991) *Coastal Zone and Open Space*, Toshibunkasha: UERG (in Japanese).

Wada, K. (1986) 'Tokyo Bay area renewal and Yasio problem', in Labour Union of Shinagawa Ward of Tokyo (ed.) *Towards a Human City: Shinagawa with Water and Green*, Tokyo: LUSWT.

8

TOKYO: WATERFRONT DEVELOPMENT AND SOCIAL NEEDS

Tetsuo Seguchi and Patrick Malone

INTRODUCTION

During the 1980s, the waterfront became a target for developers in a number of Japanese cities. While the demise of and exodus of industrial and port activities yielded large sites, Japanese development interests increased their power to exploit the waterfront through land reclamation and the creation of artificial islands. Large marine projects, which could be linked to central city areas and developed with the support of proactive metropolitan governments, held the promise of substantial development profits. In Japan's major waterfront cities, the pressures for urban growth were mobilized by public and private interests which extended cities and their commercial cores into waterfront areas.

In the late 1980s, the pressure for urban development was also spurred by the remarkable expansion of Japan's 'bubble economy'. With attention turned towards the waterfront, economic growth boosted the demand for space and the pressure to invest in property and urban development. Large marine developments signalled the expansion of Japan's urban economies and the drive for investment outlets which, for the rest of world, was evident in Japanese foreign investment in property and other media.

In Tokyo, the drive to invest in large marine projects existed alongside the problems of urban congestion, overloaded transport systems and deteriorating housing conditions. The push to expand the city into Tokyo Bay ran counter to concerns that indicated a need to limit growth and to address directly some of the city's problems. However, marine developments were promoted by the Tokyo metropolitan government and others in terms of their benefits to urban economic growth and consumption. Expansion into Tokyo Bay was also presented as a response to the city's needs and as part of the solution to Tokyo's urban problems.

It was under these circumstances that the Tokyo Waterfront Subcentre was launched in 1987. Also referred to as Tokyo Teleport Town, the project was initiated and planned by the Tokyo metropolitan government. In promoting this major development, the metropolitan government aimed to strengthen Tokyo's economic base, to release pent-up development pressure and to attract into Tokyo a share of Japan's burgeoning investment funds.

The creation of a new commercial centre in Tokyo Bay has significant implications for the commercial core and the relocation of office activities between the core and the waterfront. For critics of the project, the ambition to create this new commercial centre epitomizes the struggle between a comprehensive and socially orientated approach to urban development and the metropolitan government's essentially economic and pragmatic approach to planning. The project has provoked confrontation between those interests that are intent on solving Tokyo's urban problems and the joint forces of government and the private development lobby. In this respect, the project also highlights weaknesses inherent in the Japanese planning system, notably the tendency to embrace a blunt capitalism and the related inclination to marginalize public opinion and critical reactions to official planning proposals.

This chapter examines the nature of the conflicts generated by the Subcentre project. Lessons drawn from the project indicate some key issues for Japanese urban planning in the 1990s.

TOKYO WATERFRONT SUBCENTRE PROJECT

The Tokyo Waterfront Subcentre covers 448 ha of new and previously reclaimed land within the inner area of Tokyo Bay (Fig. 8.1). Only 90 of the Subcentre's 448 ha is private land, 338 ha is land held by the Tokyo metropolitan government and a further 80 ha will be reclaimed from the sea.

In the eyes of the metropolitan government, the Subcentre will be a high-tech, international city for the twenty-first century – a city based on advanced work practices and the global economy. Work, housing and leisure areas will be integrated in a project that will provide space for 106,000 jobs and 63,000 residents. The image projected by the metropolitan government is that of a high-tech city with a human face.

The Subcentre is divided into four districts (Aomi, Ariake South, Ariake North and Daiba)(Figs 8.2 and 8.3). A pedestrianized promenade 80 m wide and 4.1 km long will draw the different areas of the project together. This avenue will act as the main social and shopping axis and will take the form of a 'liner park' made up of walkways, squares and landscaped areas. Thus, the Subcentre's largely modernist master plan incorporates a traditional planning device in that the districts of the plan are drawn together by a heavily landscaped central axis.

The four districts of the Subcentre will contain a mixture of uses, but each district will have an individual identity based on one predominant function. Thus, the functional structure of the project can be divided into four parts which are dominated in turn by:

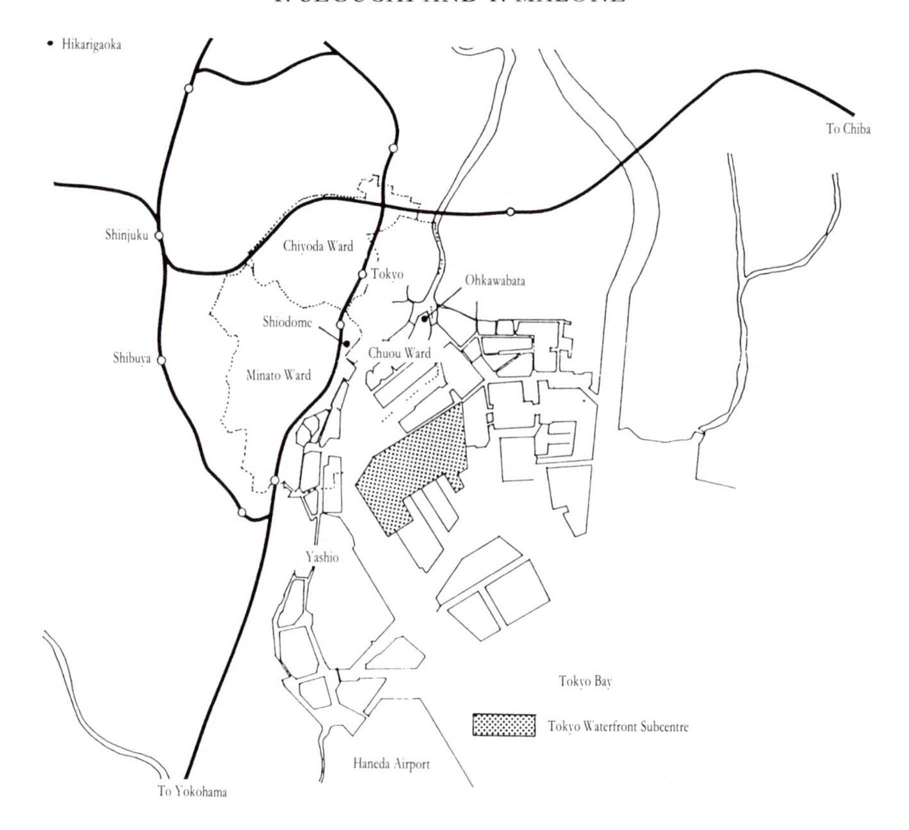

Figure 8.1 Tokyo: Waterfront Subcentre: location map

(a) commercial functions,
(b) exhibition and conference facilities,
(c) housing, and
(d) leisure facilities (Tables 8.1 and 8.2).

Arguably, the Aomi district will be the most important of the Subcentre's four districts, in that it is designated as the principal business and commercial area. It will be equipped for the latest information technology and the global economy with 'intelligent' buildings and a twenty-one-storey telecommunications centre. Its high-tech space will accommodate a work-force of 68,000. This district is central to the metropolitan government's aims to build a working '24 hour city within a city' (Fig. 8.4).

Ariake South, the second district of the Subcentre, has been described as an 'international convention park'. Together with the commercial core in the Aomi district, Ariake South shows the scale of the Subcentre project. It is expected to become Tokyo's premier international convention centre,

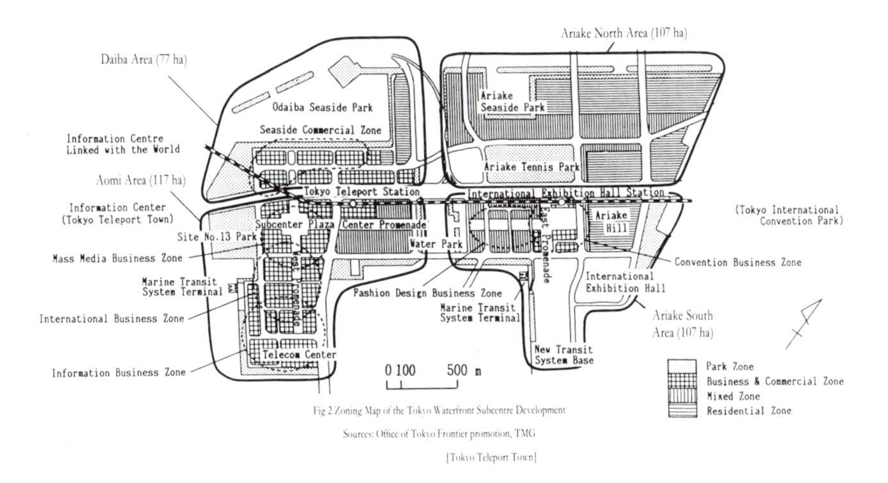

Figure 8.2 Tokyo: Waterfront Subcentre: zoning map
Source: Office of Tokyo Frontier Promotion, TMG, 1990

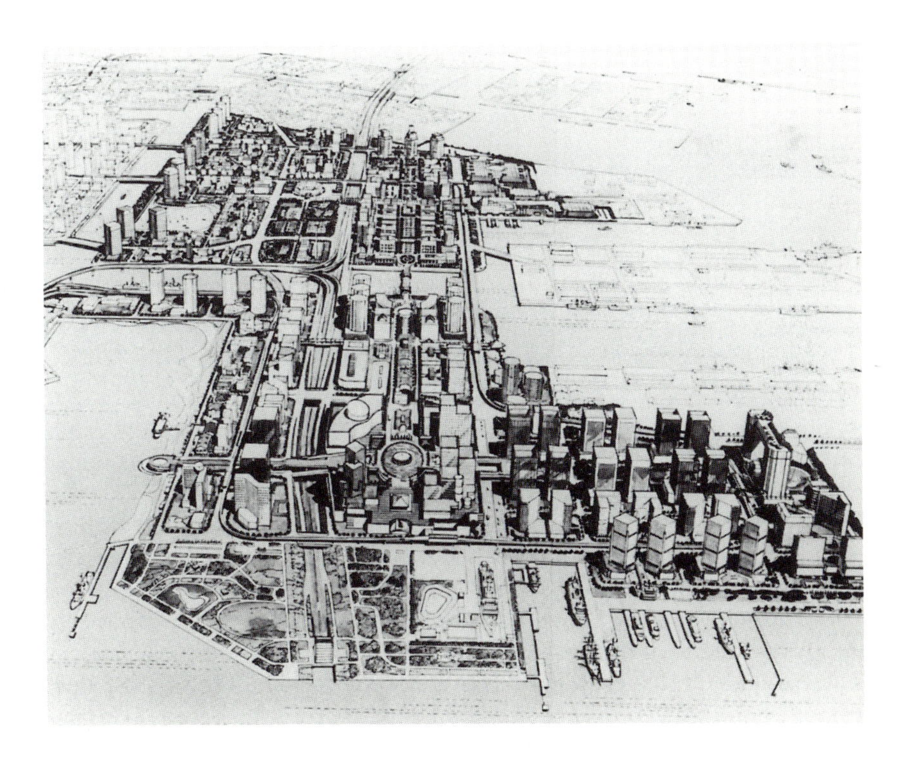

Figure 8.3 Tokyo: view of the Waterfront Subcentre
Source: Office of Tokyo Frontier Promotion

Table 8.1 Tokyo Subcentre: proposed land uses

Use	*Site area covered (ha)*	*Area of Subcentre (%)*
Roads and walkways	116	26
Main promenade	26	6
Parks	93	20
Total	235	52
Offices/retail	58	13
Mixed housing/commercial	48	11
Housing	36	8
Other functions*	71	16
Total site/building areas	213	48
Total	448	100

Source: Office of Tokyo Frontier Promotion, TMG, 1990
Note: *Includes sites used for transport facilities, civic functions and exhibition areas.

Table 8.2 Tokyo Subcentre: size and proposed populations of the four main areas

Area	*Size (ha)*	*Working population*	*Resident population*
Aomi	117	64,000	8,500
Ariake South	107	12,000	16,000
Ariake North	147	14,000	33,000
Daiba	77	16,000	5,500
Total	448	106,000	63,000

Source: Office of Tokyo Frontier Promotion, TMG, 1992

providing international convention facilities, 80,500 m^2 of exhibition space and a multipurpose hall for 2,000 people. The total floor area of the exhibition centre will be 230,873 m^2. It is also intended to promote the exhibition centre as a major venue for the fashion industry.

The third area of the Subcentre, Ariake North, will contain high-rise apartments for over 33,000 residents, over half of the proposed resident population of the Subcentre. Described by the metropolitan government as 'relaxing and appealing', Ariake North will also provide some commercial and recreational facilities, including a multipurpose sports centre.

Recreational facilities will be the major element in Daiba, the fourth district of the Subcentre. The core of this district will be a waterfront park with water-based leisure facilities (Fig. 8.5).

All four areas of the Subcentre will contain landscaped and waterfront leisure areas. Landscaping will brighten up the promenade or pedestrianized

Figure 8.4 Tokyo: Subcentre, plaza in the Aomi district
Source: TMG Tokyo Teleport Town

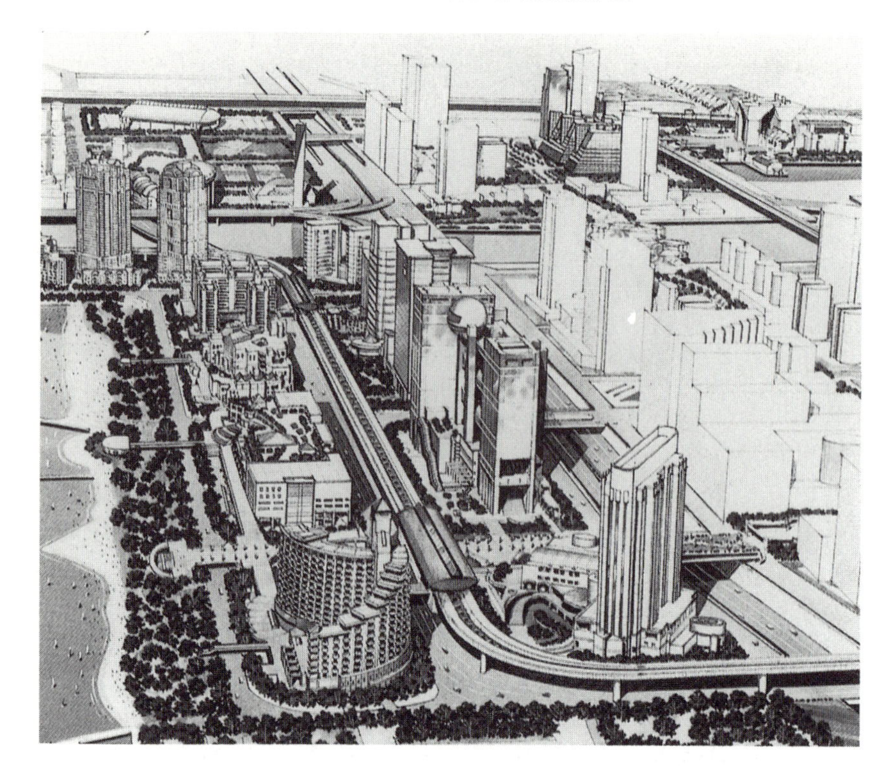

Figure 8.5 Tokyo: Waterfront Subcentre view of the Daiba district
Source: Office of Tokyo Frontier Promotion

central axis, the secondary roads and the network of pathways that will link the various parts of the project. The total area of landscaped open space will be 119 ha or 26 per cent of the area of the entire project. Hence the Tokyo metropolitan government's claim that the Subcentre will be a 'human place' set in a landscape of land and water (OPP 1988: 13–15).

For marketing purposes, the project's more obvious image is that of a metropolis for the next century. Overall, its 'point block' imagery contrasts with London Docklands and other large-scale projects of the 1980s, which eschew modernism in favour of a *beaux-arts* approach to master planning. Its architecture and planning are reminiscent of the 1960s. Thus, its futuristic architecture is already dated. Moreover, the project is a hybrid in that its ground plan leans just a little towards *beaux arts* planning. This is evident in the use of landscape, and in the open spaces of the plan which are designed to bring all the parts of the project together.

The way in which the open spaces of the project are designed to lend coherence to the master plan reflects a development process in which the plan forms the basis for the controls that the metropolitan government exercises over the various developers who contribute buildings to the project. This approach also means that the metropolitan government pursues its urban design and social values largely within the spaces between buildings and in the public spaces of the plan.

Whatever the appropriateness of its architecture and planning for the next century, the Subcentre project will be a test-bed for new technologies such as regional air conditioning, piped refuse collection, energy and water conservation, and cable TV. Its elevated pedestrian routes and waterside parks also require a sophisticated system of design and development controls and a framework for agreements between planners and private developers (Ishizawa 1987: 201–2).

Given its advanced nature, the metropolitan government planned to use the Subcentre project to show off Japanese construction and planning processes in an international exhibition. Initially scheduled for 1994, the World City Expo (originally called the Tokyo Frontier Exhibition) was to be based in Ariake South. Scheduled to coincide with the completion of the first phase of the Subcentre, this exhibition was intended to promote the project and to launch its convention and exhibition facilities. It was postponed in 1991, scaled down in 1993 and finally cancelled in May 1995 – a sequence of events that reflects the history of the Subcentre.

DEREGULATION AND THE PRIVATE SECTOR

The boom in waterfront development, which emerged in Japan in 1987, had the backing of the Japanese government. The Nakasone administration of 1987–89 launched a policy of deregulation, which created the political and institutional context for an invasion of the waterfront by development interests. Deregulation allowed the private sector to penetrate parts of the economy which, hitherto, had been the domain of the public sector. The Japanese National Railways, the Nippon Telegraph and Telephone Corporation and other public corporations were exposed to the interests of private capital. Alongside the restructuring of communications and transport, legislative changes also forced deregulation within planning and urban development.

Deregulation and rising investment pressure within the economy reinforced the invasion of planning and urban development by private capital and the invasion of the waterfront by private interests. Land-use policies were relaxed and a series of acts were passed that released specific areas of development to private interests. The Private Sector Revitalization Act (1988) gave public bodies the power to use the resources of the private sector in the construction of advanced port facilities, effectively opening up Japan's waterfronts to private investment. The act also encouraged investment in projects based on international or global 'exchange'. Given that the act targeted urban and regional development with an eye to the global economy, the term 'exchange' might be taken to mean economic trade, telecommunications, information technology, or 'cultural capital' in the form of international conference or exhibition facilities. Given too, that the aim of the act was to promote development through private-sector participation in return for tax concessions, grants, low-interest loans and public sponsorship, the act provided a basis for a sponsored invasion of the waterfront by large commercial developments orientated towards the global economy and advanced technologies. In short, the act was a recipe for the Tokyo Waterfront Subcentre.

Deregulation signalled a change in attitude towards waterfront development in Tokyo and other Japanese cities. Previously, the land use in port areas was strictly controlled and restricted to port functions, facilities for storage and transportation and allied infrastructure in the form of roads and railways. Under the Harbour Act (1950), no residential or office space could be developed in a harbour area unless that area's designation was formally annulled. Thus, in common with other ports, development in Tokyo's port area was limited to projects that exploited the power and efficiency of the port. However, in the Tokyo Metropolitan Government's Fifth Tokyo Port Plan (1988b) the Council for Ports and Harbours paved the way for re-zoning. Part of the area that will be covered by the Subcentre, plus some adjoining areas, were re-zoned. In addition, newly reclaimed land in Tokyo Bay was set to fall automatically under new zoning regulations. Thus, with the removal of planning barriers and the backing of new supportive legislation, the ground was prepared for massive public-sector investment in Tokyo Bay. More to the point, given public-sector investment in infrastructure and the removal of planning controls, conditions were created within which the private sector could generate substantial profits and feed the investment-hungry Japanese banking and financial system.

THE AMANO PROPOSAL

In 1986, the Policy Research Council of the ruling Liberal Democratic Party put forward a plan that addressed the future development of Tokyo, while focusing on the potential of the private sector. Known as the Amano Proposal, this plan was a product of political and economic ambitions rather than physical planning objectives.[1] It identified three major potential development areas in the Tokyo metropolitan area. Taken together, these three areas were expected to generate development on 427 ha of land. Approximately 400 ha of Tokyo Bay made up the largest of the three areas. The small size of the other two areas (10 ha and 17 ha) makes it clear that the Amano Proposal was directed principally at the 400 ha in Tokyo Bay (Fig. 8.6).[2]

The Amano Proposal was aimed at expanding Tokyo's economic base

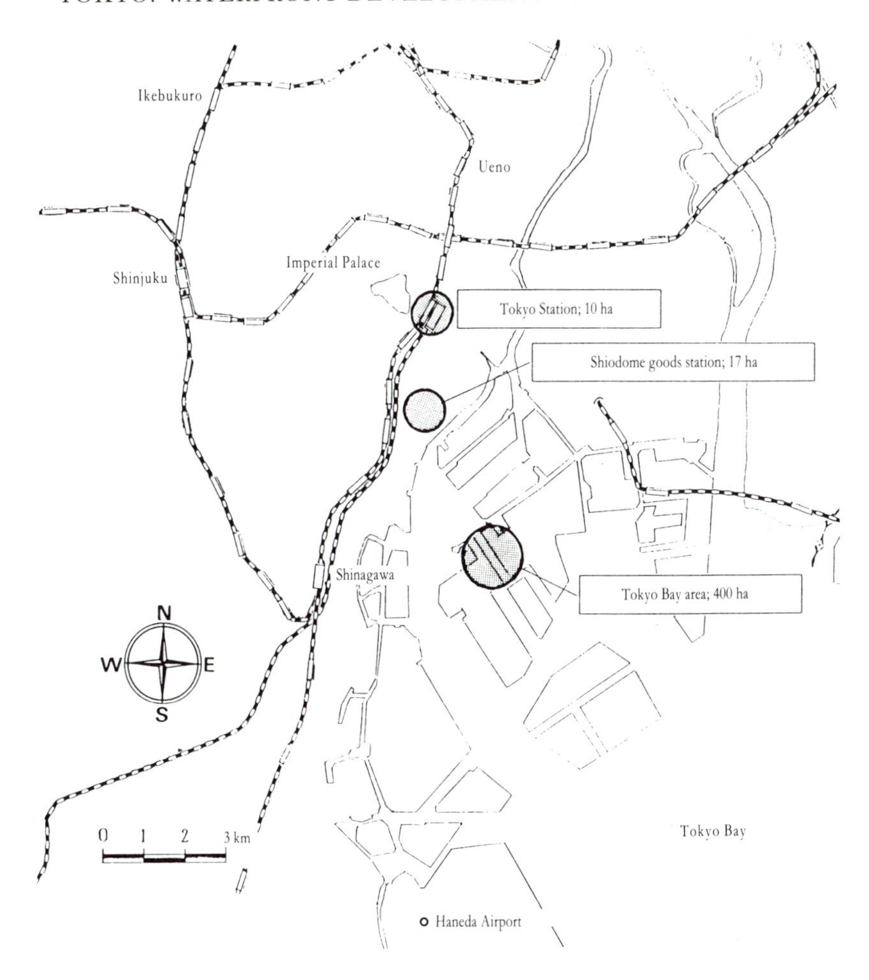

Figure 8.6 Tokyo: development areas suggested by the Amano Proposal

and the 'functional demands' of the city. It provided a framework for physical development. Acting on the Amano Proposal, the Tokyo metropolitan government, the Japan Railway Corporation (which was privatized in April 1987) and representatives from six key ministries met to chart the future of Tokyo's metropolitan area. The group, which looked specifically at the three development areas identified in the Amano Proposal, included representatives from the National Land Agency and the Ministries for Construction and International Trade and Industry.

The Environment Agency had no opportunity to participate in the meetings of the group or to express directly its opinion on the proposed projects. Moreover, local interests were generally excluded. Thus, the future of

Tokyo's metropolitan area was imposed from above by interests that supported the national government's liberalization of the economy and planning and the metropolitan government's ambitions for development in Tokyo Bay.

Given the area of land covered, and the average relationship of site to built floor area (1:3.5), the Amano Proposal allowed for the creation of 12 million m^2 of new floor space. Of this, 60 per cent was allocated for business and 25 per cent for housing. On this basis, it was estimated that development would generate 250,000 jobs and provide housing for 70,000 people.

The Amano Proposal stimulated a rash of large-scale development proposals for the Tokyo Bay area. For example, one proposal from a private group (the Tokyo Bay Research Committee) called for the construction of a new marine city on reclaimed land in the centre of Tokyo Bay, which, if built, would have provided a remarkable 45 million m^2 of office space and 80 million m^2 of residential space. This project had the capacity to create roughly 2.7 million jobs and accommodate 1.5 million residents, at an estimated cost (in 1986) of approximately 50,000 billion yen (roughly US$370 billion). This consortium also proposed further development that would have required an additional 10,000 billion yen (US$106 billion in 1995) to provide office space for an additional 323,000 workers and housing for 572,000 residents. These proposals played an important part in stimulating interest among private investors and developers in Tokyo Bay. They also focused public attention on the possibility of massive private developments on Tokyo's doorstep.

DEMAND FOR OFFICE SPACE IN TOKYO

The Amano Proposal and the ambitions of the Tokyo metropolitan government were fed by expectations of high demand for office space. In 1985, The National Land Agency (which was represented in discussions of the Amano Proposal) estimated that 86.1 million m^2 of new office space would be required in Tokyo by the year 2000. This, and other estimates provided by national agencies, provided quantitative evidence that was used to support office development proposals on the waterfront (Fig. 8.7 and Table 8.3).

Later, it became clear that some of these estimates exaggerated the potential for office development, whether intentionally or unintentionally (Ohtani 1988: 21). In the late 1980s, however, the prospect of increasing demand for office space was used by the metropolitan government to support the argument that deregulation and pent-up development pressure could produce chaotic development in central Tokyo. The case for development in Tokyo Bay was reinforced by images of massive development pressures and the difficulty of satisfying demand within the existing

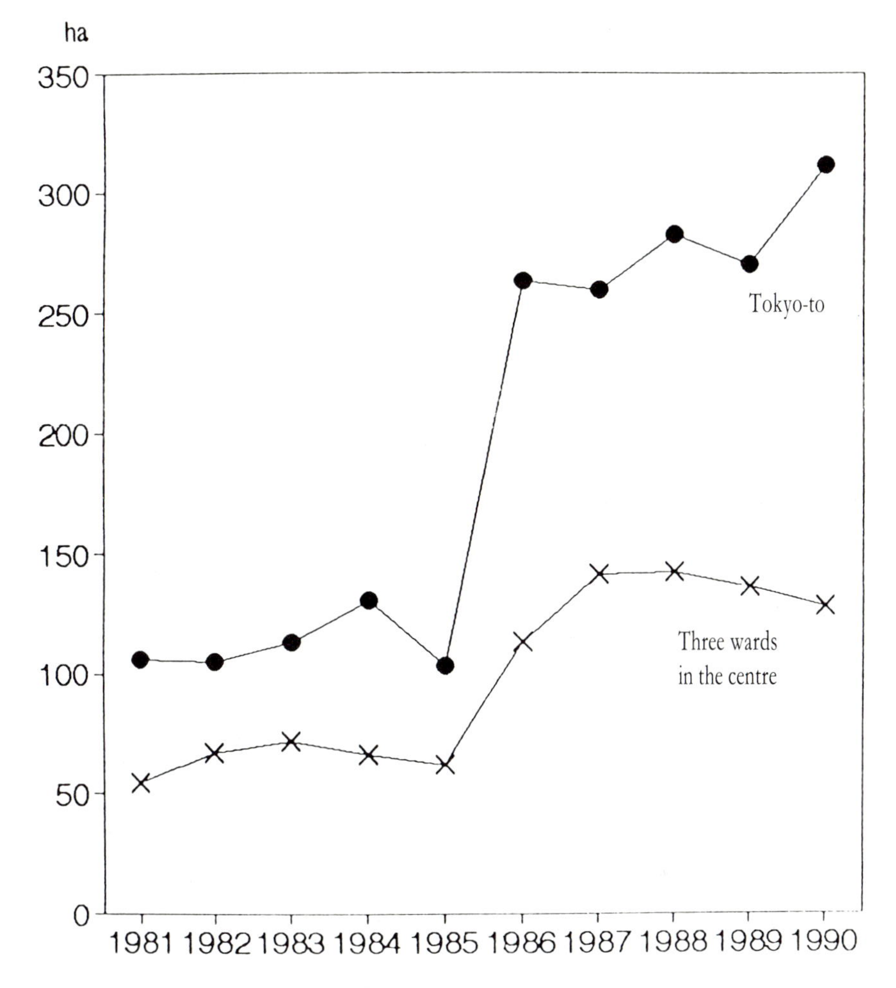

Figure 8.7 Tokyo: estimated increases in office space from 1981 to 1990
Source: TMG Land of Tokyo 1991, 1992–96

urban core. In contrast, Tokyo Bay invoked visions of an international financial and business centre and large-scale, technically advanced, urban development on land owned and controlled by the metropolitan government. Moreover, the government's case was enhanced by the weight of private interest in the redevelopment of Tokyo's waterfront.

The logic of the metropolitan government's argument was spurious insomuch as the development pressures on Tokyo Bay were generated partly by its ambition to promote Tokyo within a global economy based on information and new technology. In other words, the government fed the argument that Tokyo should expand into Tokyo Bay. Moreover, the

Table 8.3 Tokyo: estimates of demand for office space for 1986–2000

Research body	*Number of wards covered*	*Demand (10^6 m^3)*
National agencies:		
National Land Agency	23	86.1
Institute of Construction Economics	23	61.2
Tokyo Metropolitan Government	5	43.5
Tokyo Metropolitan Government	3	31.3
Credit Bank of Japan	23	57.8
Economic Planning Agency	23	60.8
Private sector:		
Mitsui Real-Estate Development Company	3	15.5

Source: Ishizawa 1987

ambition to enlarge Tokyo as a physical and economic entity ran counter to national development policies. The adoption of the Subcentre project by the Government of Japan's Fourth National Comprehensive Development Plan (1987), and other proposals for Tokyo Bay, contradict elements of national policy that call for decentralization and limits on Tokyo's economic growth.

In 1991, the National Administrative Reform Council recommended that a number of measures should be put in place to counter Tokyo's position in Japan's urban hierarchy. To this end it recommended the relocation of office functions from the Tokyo metropolitan area and the use of tax incentives to support regional planning objectives. However, the Reform Council has not had a significant influence on planning in Tokyo Bay. Neither have the metropolitan government's own analyses, which, although at odds with national policy, show the importance of offsetting centralization in Tokyo (BCP 1990: 14–19; OPP 1991: 32–7).

It is clear that the development of 6.8 million m^2 of floor space in the Subcentre project can only contribute to the centralization of economic activity in Tokyo. In this respect, both national and local governments have chosen to ignore the decentralization issue in Japanese regional policy. In the case of the Subcentre project, the metropolitan government is using public land and public funds to frustrate decentralization policies.

ECONOMICS AND THE DEVELOPMENT PROCESS

Direct investment by the metropolitan government in the Subcentre project until 1990 was estimated at 3,960 billion yen (roughly US$42 billion). This covers the estimated cost of regional transport facilities and main roads (2,790 billion yen), and the preparation of land and urban infrastructure

(1,170 billion yen). The metropolitan government intends to recover its expenditure on the project and the cost of any associated infrastructure, principally through the use of long-term (sixty-year) leasing agreements. It also interprets the direct involvement of the private-sector in planning and design as a means of capitalizing on private-sector resources and thereby reducing public expenditure.

Given that the development process employed by the metropolitan government relies on the private sector, the first invitation to developers to submit competitive proposals was launched in May 1990. This invitation covered 23 ha of the Subcentre. A total of eighty-three proposals were submitted and fifteen were selected for further consideration.

Although the collusion between private and public interests in the Subcentre project is open to criticism, the metropolitan government does exert a plethora of development and other controls over the private sector. For example, development leases and public land ownership guard against the dramatic increases in property prices that are characteristic of Tokyo's private property markets. Under this policy of public land ownership, however, powerful developers may try to enhance development profits by increasing densities for commercial and housing developments. This method of increasing profits is permitted for developers because the ground rents that developers return to the metropolitan government are fixed in relation to land values in contiguous central city areas and not on the basis of potential or actual land use, densities or rents. Thus, developers are encouraged to maximize rents through the manipulation of land uses and through high-density development. The absence of a direct relationship between ground rents and development profits also exacerbates doubts about the level of ground rents charged by the metropolitan government. Moreover, whereas ground rents are generally set at 2.5–3 per cent of the land's value, critics claim that the assumed land values may be roughly half those of real market values (and in one case as low as 1/15 of the market value)(Misawa 1991: 48).

In the face of the argument that ground rents should be based on development profits, and should distinguish between land used for housing and land covered by commercial functions, the government clings to its strategy of long-term 'developer-friendly' leases and long-term recovery of public expenditure. This is because its policies on land and the retrieval of public expenditure support its ambitions to attract private investment into urban development. Its method of calculating land rents is, effectively, an incentive scheme, which could be compared to alternative strategies based, for example, on the remission of taxes or rent allowances. Moreover, this type of incentive scheme favours commercial forms of development.

CRITICAL REACTIONS

From the outset, the government's efforts to advance the Subcentre have provoked controversy. Communities and labour unions organized study groups to discuss the project and made their dissent public (Misawa 1991: 56–81). Academics made alternative proposals focusing on doubts regarding the environmental and housing aspects of the project. Some of the city's twenty-three wards have made assessments of the project that are at odds with those of the metropolitan government. The inhabitants of Koutou Ward, for example, drew up their own reports and appealed to the government to reassess its plans on the grounds of the potential impact of the project on the quality of urban life (Koutou Ward 1990).

Criticism has been levelled at the project itself and its significance for Tokyo. The basic question is whether this, and other developments in Tokyo Bay, will aid or hinder the solution of Tokyo's urban problems. The conflicts raised by the Subcentre, which have polarized the citizens of Tokyo and the city's administrators, reflect the struggle between those who wish to resolve Tokyo's problems and those who have ambitions to expand the city and its economy. In general terms, these conflicts focus on environmental concerns, traffic problems and housing issues.

For many of its critics, the Subcentre raises substantial environmental issues. The counter-proposals that support the 'greening' of Tokyo Bay demonstrate that there is a large gap between the metropolitan government and its critics in terms of the provision of recreational areas and the perception of environmental and ecological issues. Alternative proposals, generated by academics and community groups, reflect the view that the government is driven by economic ambitions and the urge for large-scale development. They argue that the Subcentre project shows a lack of consideration for Tokyo's urban problems, for the needs of its citizens and for the environmental capacity of the city and Tokyo Bay.

One of the major areas of contention is the use of Tokyo Bay for predominantly commercial rather than recreational developments. The total area of urban parks and 'green' open spaces in Tokyo in 1994 (5,608 ha) amounted to only 4.8 m^2 per head of population. That is roughly half the standard of 10 m^2 per person set down by Japanese legislation as a target for the year 2000. In theory, the Tokyo metropolitan government is legally bound to provide more open space in the city. However, rising land prices make the acquisition of land increasingly difficult and, if the population of Tokyo continues to increase, the metropolitan government may not even maintain the present low standard of provision. However, in opposing its critics, the metropolitan government has allied the achievement of open space targets to the redevelopment of Tokyo's river and waterfront areas and has used the issue of open space to support development in Tokyo Bay.

Planners within the metropolitan government accept the importance of the waterfront in terms of the creation of parks and open space (Moroboshi 1992: 69–70). While critics argue that a substantially greater area of the waterfront should be used for parkland rather than for commercial development, the metropolitan government holds that many waterfront parks have already been built to relieve the shortage of public open space. However, the integrity of the government's response can be measured in terms of the relatively low ratio of parkland in Tokyo.

Oshima has proposed a dramatic reduction in the Subcentre project in favour of environmental conservation in Tokyo Bay (Oshima 1991: 10–12). In opposition to the metropolitan government's plan for the Subcentre,

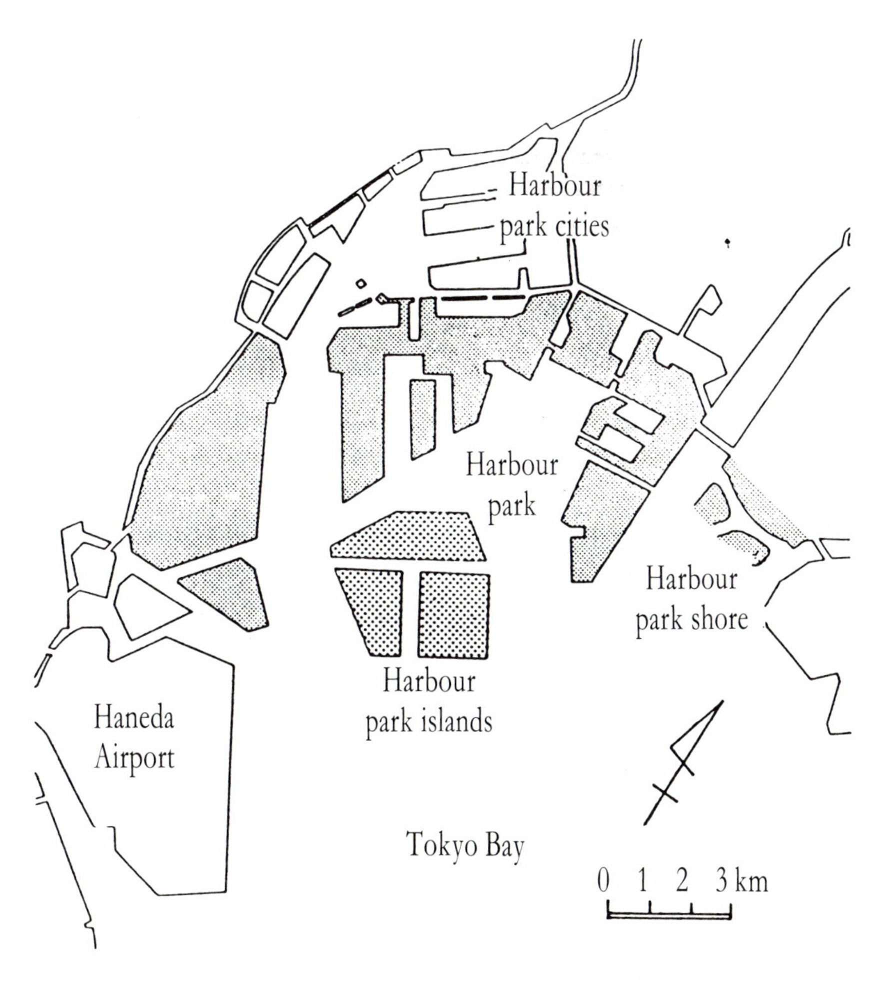

Figure 8.8 Tokyo: Central Park proposal for Tokyo Bay
Source: Tajiri 1988

Oshima argues that the construction of two–five-storey apartment blocks, allied to a policy of environmental protection, would create a satisfactory living environment and provide areas for recreation and wildlife. It is also argued that land reclamation work on the Subcentre project should be stopped, and that tidal areas should be used for sea-water purification and as a habitat for bird and marine life. Thus, a process of natural recovery in Tokyo Bay might allow for the creation of a network of new and existing parks, and for the provision of an 'urban forest' – a form of green open space that Tokyo lacks.

Another proposal, which has also emerged from academic circles, favours the construction of a 'harbour park', built on 28 km^2 of artificial islands, to serve the Tokyo metropolitan area (Tajiri 1988)(Fig. 8.8).

TRANSPORT

Although only one of a number of recent developments in Tokyo Bay, the Subcentre requires considerable expansion in Tokyo's transport infrastructure in the form of a complex network of new motorway and rail links (Figs 8.9, 8.10 and Table 8.4).[3]

The extent of the public transport infrastructure required to support the Subcentre is indicated by its cost which, in 1990, was estimated at about 520 billion yen (roughly US$5.5 billion). The impact of the Subcentre on Tokyo's transport networks might also be assessed in terms of its size and its projected work-force of 106,000 people. It is estimated to generate a total of roughly 450,000 outward and return trips daily. Thus, it is doubtful that the proposed developments in Tokyo Bay will alleviate the city's traffic problems. Rather, they are likely to exacerbate the difficulties of a city that is already suffering from congestion and overloaded transport systems. Moreover, given that new roads constructed to relieve congestion in Tokyo are soon jammed with traffic, the same fate is likely to befall the roads that are being built to serve the Subcentre. In addition, the traffic generated by the project will have an adverse effect on air quality.

The environmental problems raised by critics of the Subcentre are not

Table 8.4 Tokyo Subcentre: estimate of journeys (units 1,000 journeys/day)

	1985	*2000*	*2005*
Car	367	1,051	1,066
Truck	421	602	618
Totals	788	1,653	1,684

Source: Bureau of Environmental Protection 1991

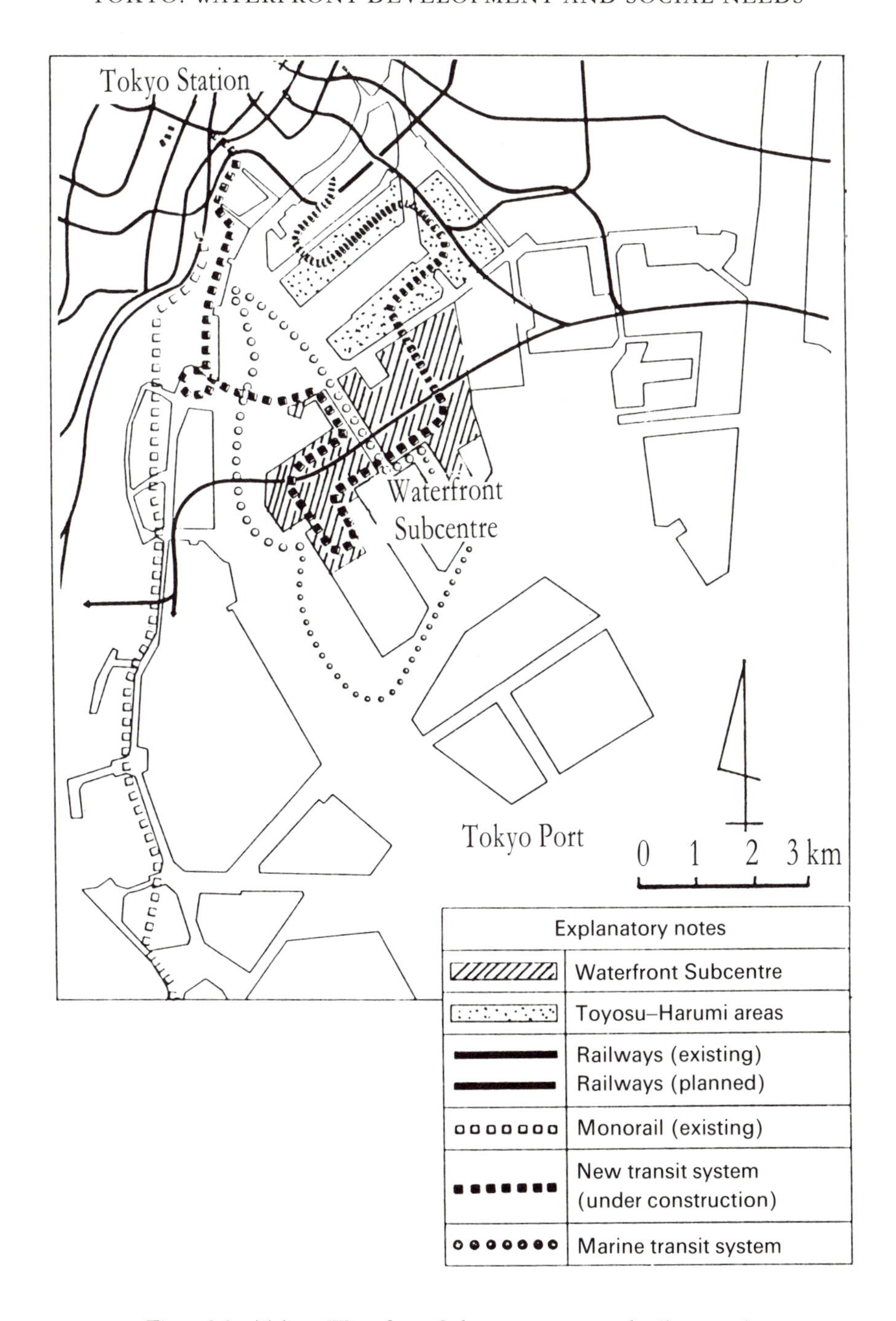

Figure 8.9 Tokyo: Waterfront Subcentre: proposed rail network
Source: Office of Tokyo Frontier Promotion 1990

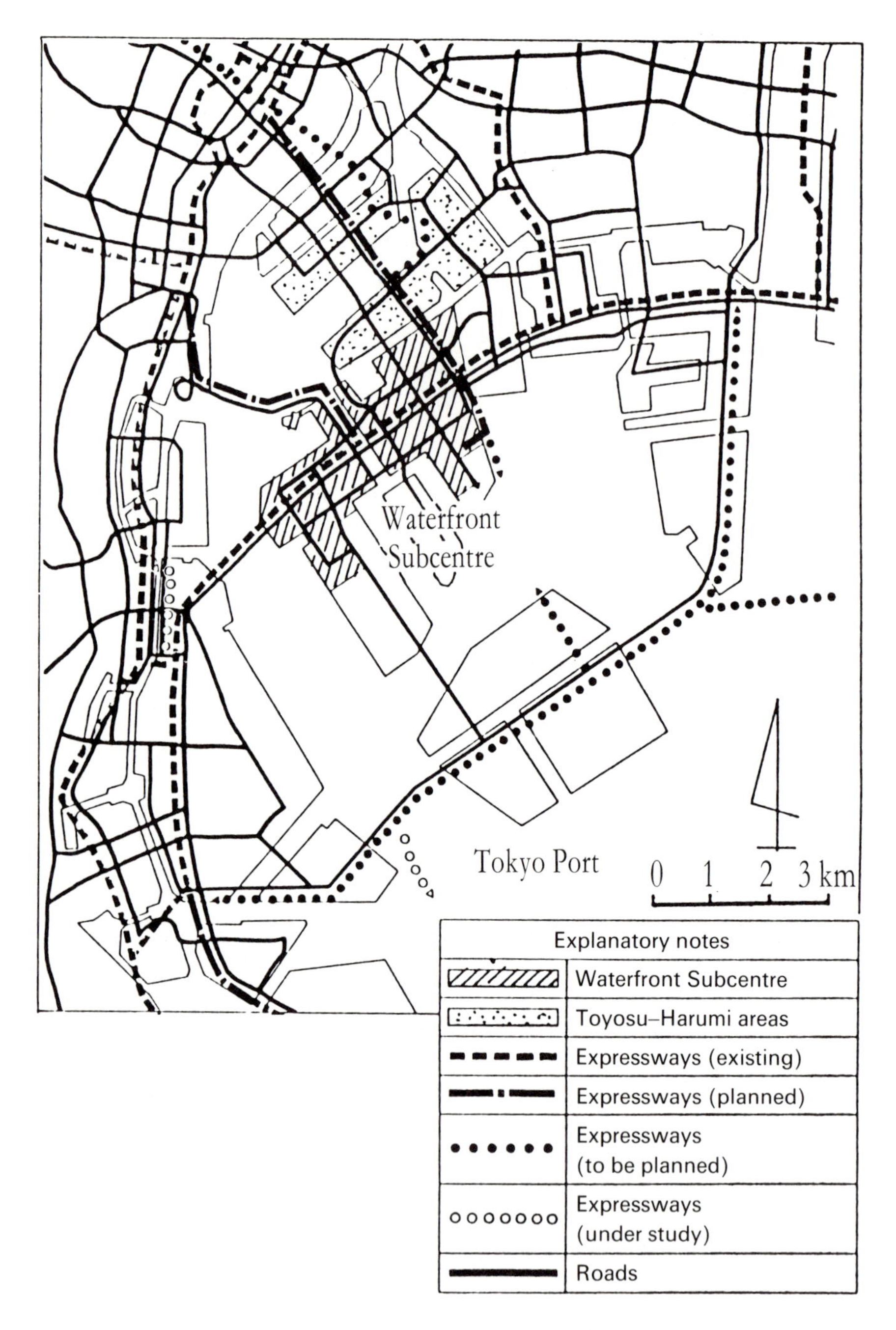

Figure 8.10 Tokyo: Waterfront Subcentre: proposed road network
Source: Office of Tokyo Frontier Promotion 1990

highlighted in the assessments made by the metropolitan government, which cover the possible effects of new trunk roads serving the project. Official studies undertaken in 1990 suggest that environmental impacts will not be severe (Bureau of City Planning 1991). Similarly, other factors that have been subject to environmental impact analyses, such as air and water pollution, have been favourably assessed. It is important to note, however, that the official analyses of these factors cover the environmental effects of individual parts of the Subcentre. The combined effects of all parts of the project have not been calculated.

In 1991, the metropolitan government conducted an environmental review that covered the Tokyo metropolitan area (Bureau of Environmental Protection 1991). The results of this assessment indicate that, overall, further development in the three main wards will not have a detrimental effect on environmental conditions in the metropolitan area. This assessment covers the proposed Subcentre, but does not provide an individual environmental 'balance sheet' for this or any other project. Thus, by failing to deal with all aspects of the project, or by lumping the project into a general environmental study, the environmental impact of the Subcentre has been obscured.

HOUSING ISSUES

Housing issues provide another area of conflict for the metropolitan government and its critics. Given the nature of Tokyo's housing problems, a more enlightened planning regime could use Tokyo Bay to relieve housing congestion and to absorb people displaced by inner-city redevelopment (Ohtani 1991: 6–7). The government has entertained this concept in the past, but Tokyo's housing problems have not had a substantial influence on its plans for the Subcentre.[4]

There are several indicators of the low standards of Tokyo's housing stock. On average, dwellings are relatively small. In 1988, the average size of housing units in Tokyo was given as 58.5 m^2. That is roughly two-thirds of the national average (86 m^2). Thus, space standards are low in comparison with some other metropolitan areas such as Nagoya (76.5 m^2), Kitakyushu and Kyoto (both 72 m^2). Almost 18 per cent of the housing stock in the city's Tokyo Ward is below minimum dwelling standards set by the Ministry of Construction. Furthermore, 86 per cent of the households which are below standard pay relatively high rents in relation to housing quality.

The metropolitan government prefers not to use public land (or Tokyo Bay) to solve Tokyo's housing problems. Instead it favours high-rise solutions and the rebuilding of existing low-rise public housing to increase densities (BCCA 1990: 75–6). This reflects a general trend. High-rise housing did not appear in Tokyo until the 1970s, but the recent completion of a forty-storey apartment block (at Ohkawabata River City) attests to its popularity as a solution for the city's housing problems – a solution that is being pursued in Japan regardless of doubts about the quality of life in high-rise apartments (Hayakawa 1987: 86–8).

Initially, 20,000 housing units were to be provided in the Subcentre and, officially, the metropolitan government has adopted the policy of providing dwellings in relation to the number of jobs in the project (BCP 1990: 7). However, the proposed number of housing units accounts for roughly half of the potential working population in the Subcentre. Thus, the number of dwellings will not be equal to the number of jobs created by the project, nor will the project relieve Tokyo's wider housing problems.

The Tokyo Metropolitan Workers' Union has appealed for a general change in housing policy in Tokyo Bay (Misawa 1991: 146–53). The Union holds that developments should be dedicated principally to improving living conditions in Tokyo. It has argued for the construction of 40,000 units of public rented housing in the Subcentre, that is, twice the number proposed by the metropolitan government. The Union also argues that the working population in the Subcentre should be limited to 30,000 or less, and that the remaining land should be used for parks and cultural facilities.

The nature of the housing proposed for the Subcentre is another area of major concern. In order to boost the capacity of the housing element while limiting the area of valuable land available for housing, the metropolitan government stipulated the construction of high-rise apartments (OPP 1989: 25–7). The highest residential blocks in the Subcentre will be thirty-five storeys (140 m), although building heights may be reduced on the basis of factors such as overshadowing (Table 8.5).

Whereas all of the housing in the Subcentre will be rented, 65 per cent will be leased as public housing by the metropolitan government, the Housing and Development Corporation (HDCO) and other institutions. The remaining 35 per cent will be private rented housing on land assigned to

Table 8.5 Tokyo Subcentre: comparative housing and population densities

Location	*Persons per ha*	*Housing units per ha*	*Net floor area ratio*
Hikarigaoka	350	100	220
Shinagawa, Yasio	450	130	180
Ohkawabata	560	160	470
Subcentre	530	180	400

Source: Koutou Ward 1990
Notes: Persons/ha or housing units/ha is total population/number of housing units, divided into the combined area of housing sites, parks and district roads (trunk roads are excluded). The net floor area ratio is the total housing floor area divided by the total area of the housing lots.

developers on thirty-year renewable leases. By retaining land ownership, the government obstructs property speculation on the part of occupiers or owners who might sell dwellings at a profit.

The provision of a relatively high proportion of public housing is intended to provide wider access to housing that might otherwise be subject to Tokyo's soaring property market. However, rent levels in the Subcentre will be high if compared, for example, with average HDCO public sector rents (Misawa 1991: 62–4). In 1991, the average rent for an HDCO apartment of 95 m^2, occupied by a family of four was 200,000 yen per month (roughly US$2,130). The cost of private rented apartments was roughly twice that of a comparable HDCO dwelling, although even public-sector rents are high in relation to average incomes in Tokyo (roughly 7.5 million yen per annum in 1991).

Within the Subcentre, a family of four living on the average wage might pay 2.4 million yen or 32 per cent of its annual income in rent for public housing. The same family might pay 63 per cent of its annual income for an apartment in the private sector, where rents are ultimately determined by the market. This prompts questions as to why rents for public housing, built on public land, can be so high. It also offers little hope that the type of housing policy adopted for the Subcentre project can do much to solve Tokyo's housing problems.

Housing policy in the Subcentre project also prompts questions of public safety. Given that artificial islands and reclaimed land generally provide poor ground conditions, and that Tokyo Bay is subject to earthquakes, 140-m-high apartment blocks conjure up visions of disaster. It must also be expected that the concentration of new developments in Tokyo Bay will compound the risks and may exacerbate the damaging effects of earthquakes (BPH 1990: 7–18). Although the Subcentre project will incorporate protective measures, the government cannot entirely assuage public doubts which are fed by anxieties regarding the additional risk of soil liquefaction, subsidence, tidal waves and damage to transport links in the event of an earthquake – anxieties that were heightened by the Great Hanshin Earthquake which led to severe damage and loss of life in Kobe and Awaji in January 1995 (see Chapter seven).[5]

POLICY CONCERNS

The debates surrounding the government's proposals for Tokyo Bay reflect deep disquiet about the general drift towards pragmatic planning. There has been a perceptible shift in public policy. For example, in the 1970s the metropolitan government adopted a 'green policy' and, partly in response to the demands of a community group, initiated the Kaijou Waterfront Park in Tokyo Bay (Tajiri 1988: 137–79). In 1988, however, when 1,057 ha of the park had been developed, the government switched to a policy of private-sector development. Thus, negative reactions to the Subcentre are underpinned by wider concerns regarding privatization and the difficulty of prising social and environmental gains from profit-orientated development processes. These concerns have been sharpened by the realization that the nature of urban development in Japan will be determined by the way in which the policy of deregulation is exercised in the middle and late 1990s. There is a tendency for the debate to focus on specific issues, for example on housing, transport or the need for public open space. It might be argued, however, that these issues are symptoms of a particular approach to development, which embodies a general tendency to neglect the public interest and to restrict the pursuit of social goals in favour of private economic ambitions; albeit that these are presented as official planning objectives.

Given the shift towards deregulation, some critics of the Japanese planning system hold that social goals and the public interest can only be served where urban development is led by public rather than private interests. This has led some critics to urge the Tokyo metropolitan government to scrap proposals for the Subcentre and to start again with a clean slate and a socialized planning system (Oshima 1991: 10–12). It is argued that the citizens of Tokyo want the government to use their taxes to create a satisfactory and socially orientated metropolis, not to exacerbate Tokyo's problems (Secretariat of Tokyo Metropolitan Assembly 1987: 221–32).

Doubts raised by the collusion between public and private interests in Tokyo Bay are accentuated by the knowledge that roughly 75 per cent of the area set aside for the Subcentre is public land, held by the metropolitan government, which is being used to accommodate private development interests rather than to address Tokyo's urban problems (Ohtani 1991: 6–7).[6] To quell these doubts, the metropolitan government argues that it can mitigate Tokyo's problems by building a 'desirable' city in Tokyo Bay (OPP 1988: 3). It says that the Subcentre project, although largely an exercise in private enterprise, will be under its control – an argument that raises fundamental questions about Japan's power structure and the neutrality of the government in terms of the interests of private capital.

To avoid the criticism that Tokyo's problems have been marginalized, the metropolitan government has attempted to portray the Subcentre project as an opportunity to address problems, particularly in the central city area. A key mechanism in this strategy is the potential to move office space from the central area to Tokyo Bay. The government has adopted the strategy of asking developers to make proposals based on linking the redevelopment of privately owned city-centre land to the development of the Subcentre (TMG 1989: 2). Effectively, developers compete to swap either land and/or development rights on inner-city land for access to the Subcentre project. The redirection of office development pressures from the city centre to the Subcentre is meant to allow inner-city land (hitherto controlled by private developers), to be used to provide public open space, housing, roads, cultural or social facilities or other uses which would meet the needs of the city centre. Individual agreements between developers and the metropolitan government may be based on 'planning gains' and/or the transfer of land ownership. In short, developers are invited to suggest ways in which property and development rights may be traded for access to the Subcentre. One development interest, for example, has suggested that a city-centre site should be transferred to the Tokyo metropolitan government at a 'reasonable' price, in exchange for access to the Subcentre. Other developers have suggested that city-centre sites could be used for housing, open space, roads and the construction of cultural facilities.

In future, the Subcentre project may provide an opportunity to examine the effects of a policy that allows local authorities to manipulate development pressures within the urban spatial structure by swapping development rights between two urban areas. It is too early, however, to assess accurately the effectiveness of this policy. Moreover, it is clear that the policy, and the potential planning gains in the city-centre, should not be allowed to obscure the issues raised directly by the Subcentre and by its potential impact on Tokyo.

PLANNING RITUALS

The Subcentre is a pragmatic project driven by economic and political ambitions rather than planning goals. It is intended to enhance Tokyo's economy and to create large profits for the private sector. However, the economic nucleus of the project is enclosed in a shell of administrative procedures. These procedures obscure the nature of the project and deflect or absorb critical reactions. The project has passed through the formal processes of authorization. However, despite adverse public opinion, negative environmental impact analyses and alternative planning proposals, the project remains fundamentally unchanged. In this respect, the shell of administrative procedures and planning rituals that surround the project act as a defensive wall that can be manipulated to address the social conflicts and political tensions raised by the project.

To understand the Subcentre project, it is necessary to examine the economic and political interests behind it, and the processes by which the aims of these interests are formalized within the public realm. The formalization process can be traced to 1987 when the metropolitan government launched the Subcentre project as a development concept. Although the initial announcement and the development plan (which followed in 1988) drew negative reactions from the public, it was clear from the outset that the metropolitan government was not keen to alter its proposals. Neither was it willing to involve the public in consultations outside those required under planning law. However, the project was passed through the formal channels.

Public and local-authority interests were afforded their statutory rights to voice opinions. The major roadworks required by the project were officially sanctioned by the Tokyo Metropolitan City Planning Council and the project was passed through the required environmental assessments. Representations were made by the leaders of city wards and residents' groups, and the project was then sanctioned by the Planning Council and later approved by the Minister of Construction.

In autumn 1991, an assessment of the environmental impact of the main trunk roads was published. However, by that time the metropolitan government had already allocated almost half of the public investment in the project (roughly 2,000 billion yen or US$21 billion) to cover expenditure up to the year 2000. This sequence of events demonstrates the emphasis given to expediting the development.

The size of the government's financial commitment secured the project as a major element in long-term plans for Tokyo. However, the groundswell of critical reactions from labour and community groups, academics and others came to a head in 1991 with the election for the governorship of Tokyo. The proposed budget for the Subcentre came under attack from the Japanese Socialist Party and other opposition groups, and, in the lead-up to the election, the budget was rejected by the Tokyo Metropolitan Assembly. The main objections raised by the Assembly concerned the number of housing units and the reckless pace of planning and development procedures. Following the rejection of the budget, construction work on the project was halted for four months. Associated infrastructure projects and negotiations between the metropolitan government and private developers came to a standstill. Thus, the project was virtually suspended for the period of the 1991 election. At the same time, the ruling Liberal Democratic Party split into two factions. One faction supported the incumbent governor (Shunichi Suzuki), who promoted the Subcentre project. The other faction supported a new candidate with an opposing view. The project became an election issue, and eventually a vote was passed in the Assembly to establish a review of the project in June 1991.

The review panel canvassed opinions on the project from community groups, academics and city wards affected by the project. The heads of Minato and Koutou wards called on the metropolitan government to reconsider the overall development framework for the project, particularly in terms of housing policy and the utilization of city centre sites that might be vacated by offices moving to the Subcentre. The Socialist Party also demanded an increase in the provision of social housing in the Subcentre. In addition, citizen groups organized symposia on the Subcentre in July and September 1991.

When the governor was re-elected in April 1991, those interests that opposed the Subcentre project and the metropolitan government's approach to development in Tokyo Bay lost a major opportunity to alter

Table 8.6 Tokyo Subcentre: main points arising from a review of the project

Housing:	(a) increase of roughly 1,000 units to 21,000 (b) 1,850 additional units by public sector (c) 60 per cent in Daiba district by TMG
Schedule:	(a) two years added to the development schedule (b) initial stage postponed until 1995 (c) World City Expo postponed to March 1996

Note: Other factors considered were project competitions, environmental issues, facilities for culture and sports and measures to counteract the effects of earthquakes.

the direction of planning in Tokyo. Following the election, the governor negotiated a political agreement on the Subcentre project with opposition parties (including the Japanese Socialist Party). He agreed to change the development schedule and to increase the number of public housing units. Consequently, a new budget was passed by the Metropolitan Assembly. However, the outcome of the debate that had been forced by the election was limited (Table 8.6).

The review panel published its draft report in August 1991. In November 1991, only five months after it was established, it published its final report. Whereas the panel made no major changes, it conceded some small gains to critics of the Subcentre project. The number of housing units was increased from the original 20,000 to 21,000. There was no significant alteration in the ratio of public to private housing, but in response to demands from the Minato Ward all the housing units to be provided in the Subcentre's Daiba district were reallocated as public housing and the number of units in that district was increased from 1,700 to 1,850 units. To allow more time for planning, the schedule for the project was extended by up to three years. The World City Expo (initially the Tokyo Frontier Exhibition), which was originally scheduled for 1993, was postponed for two years amid doubts about its content and purpose.

The review panel recognized the potential for using land vacated by offices relocating to the waterfront to help solve Tokyo's urban problems. It supported the use of central city land for infrastructure, housing, parks, open space and social welfare facilities. It also recommended that the potential of vacated sites in the city centre should be subject to additional research.

Other matters raised by the review panel included, for example, the dissemination of information regarding the Subcentre project, the nature of competitions, environmental issues and the threat of natural disasters. Sadly, however, the review was politically constrained in terms of its scope and the power of the panel to make decisions.

The outcome of the review demonstrated how the formal processes and planning rituals that 'fronted' the Subcentre project could absorb attacks from critics while, essentially, the proposal remained intact. The project survived only to be buffeted by the economic recession which, after 1991, created conditions characterized by faltering investment, revised contracts, reduced rents and rising public expenditure.

In 1995, Yukio Aoshima replaced Shunichi Suzuki as governor of Tokyo. In Aoshima's election campaign he promised to review the Subcentre project and to cancel the proposed World City Expo – a promise which he fulfilled. This cancellation reflects the deepening recession and the implosion of Japan's 'bubble economy'. It might also be interpreted as evidence of a phenomenon whereby large waterfront developments, pioneered by politicians in the 1980s, are suppressed by their political successors in the 1990s. Waterfront projects that were politically 'bankable' in the expanding economies of the 1980s may now have a new political value. The suppression of projects may become part of the electoral platform of politicians who must enforce economic stringency in a recession, and who may depict political opponents and the projects that they promoted in the 1980s as profligate and imprudent.

In newspaper interviews regarding the decision of Toyko's new governor to cancel the World City Expo, one citizen remarked that Aoshima 'represents Tokyo citizens' opinions better than the Tokyo assembly members'; another stated that 'the person to blame is the one who started it' (*Japan Times*, 2 June 1995: 1). These comments may reflect the attitude of many of Tokyo's citizens towards the curtailment of future expenditure on the Subcentre and comparable projects. However, Japan has yet to count the full costs of its recession, and to assess fully the role of property and urban development in the 'bubble economy'. In this respect, Tokyo's Subcentre project now exists against the background of major crises in Japan's banking system, the failure of the country's largest credit union and a total volume of bad loans estimated at 60,000 billion to 100,000 billion yen (roughly US$600 billion to 1,000 billion).[7]

CONCLUSIONS

Critics argue that the Tokyo Waterfront Subcentre project is driven by economic and pragmatic goals. It awards priority to private economic ambitions rather than to public needs or the solution of Tokyo's urban problems. However, the project is not unusual, given the pragmatic nature of Japanese urban development. It demonstrates a typical reluctance to come to grips with the possible impact of large-scale developments on the environment and on existing urban problems. It shows how planning may embody environmental impact analyses, or conform to legal requirements regarding public consultation, and yet marginalize environmental and social issues.

Public concern about environmental and planning issues, and a major official review, have produced only minor changes to the Subcentre project. This demonstrates that those who criticize urban projects in Japan, or who make demands of urban governments, are in a weak position. Japanese planning legislation lends a great deal of power to public bodies that are not always responsive to public needs. Given that Shunichi Suzuki, as governor of Tokyo (April 1979–April 1995), drew political support from development interests, it was perhaps unrealistic to expect that public opinion might force fundamental changes in the Subcentre project. Given the context for development, public consultation, although mandatory under planning law, is unlikely to shift the balance of power in favour of the public interest. Ultimately, decision-making is in the hands of political and economic interests that may ignore dissenting voices. There is no legal framework that might force a referendum on the Subcentre, and public participation in the planning process is given little value. The review panel that examined the Subcentre project was made up solely of staff from the office of the metropolitan government that is responsible for the project. Thus, the government sat in judgement on itself. Moreover, the governor of Tokyo retained the right to ratify even the minor changes that might be suggested by the review panel.

The restrictions placed on public participation, on environmental assessments and on the capacity of either research or public opinion to overturn planning decisions are major faults in Japanese planning. They indicate a planning process that is geared to economic and political goals and an approach to planning which can carry significant social and physical costs. Moreover, since its emergence in the heyday of Japan's 'bubble economy', the costs of the Subcentre have risen with the onset of an economic recession. Regardless of the degree of official support and state sponsorship, the Subcentre project is subject to underlying fluctuations in the economy and property markets. Thus, the conditions surrounding the project changed markedly with the onset of the economic recession in mid-1991. Negotiations between the metropolitan government and developers foundered. The government was forced to revise contracts with developers and reduce rents by 36 per cent. While the election for governor and the subsequent review of the project took place against a background of economic change, some of the private interests associated with the Subcentre began to demand alterations in contracts relating, for example, to ground rents and contract prices. Developers asked for permission to increase previously agreed office floor areas to compensate for falling rents. Ultimately, the metropolitan government was forced to compromise and to increase public expenditure on the project by about 100 billion yen to cover reductions in ground rents and premiums. It is expected that other compromises will follow, and with each setback in the Subcentre project the metropolitan government's alliance with the private sector will become more suspect.

The gap between the project and the aspirations of the public is also widening as it becomes clear that initial predictions of the profits for the metropolitan government were based on land values in a booming economy. Today, as hopes of commercial success fade, the metropolitan government and private interests are facing greater difficulty in their efforts to promote the project. The difficulties that now face Japan's economic system, and the prospect of further collapses in a banking system burdened by bad loans, may lead to a reassessment of the Subcentre and other projects in terms of their relevance to Japan's economic problems. Valuable lessons can be drawn from the failure of Japan's 'bubble economy', but this is cold comfort for those who harbour doubts about the social value of the Subcentre project and its implications for planning in Japan.

NOTES

1. M. Amano was a politician associated with the Ministry of Construction (July 1986–November 1987) and a member of the House of Representatives and of the ruling Liberal Democratic Party.
2. The two other areas covered by the Amano Proposal are 17 ha on the site of the former Shiodome goods station and 10 ha around Tokyo Station.
3. According to the metropolitan government's plan the project will require the construction of Loop Road no. 2, the New Tokyo Waterfront Transit System, the Maritime Transport System, the Urban Expressway Harumi Line and Expressway no. 12.
4. Clearly, critics of the Subcentre may promote conflicting uses for the waterfront, in terms of open space and housing. In reality, however, open space and housing have been given such a low priority that the waterfront is unlikely to provide a basis for the solution of Tokyo's problems.
5. The metropolitan government now favours the use of ramped embankments (as opposed to humped embankments, which prohibit access to green areas near the water).
6. The area of the Subcentre is made up of 338 ha of land held by the Tokyo metropolitan government, 90 ha of private land and 80 ha of reclaimed land.
7. The Ministry of Finance estimates bad loans at 50,000 billion yen. Outside analysts put the figure at 60,000–100,000 billion yen (*Guardian*, 31 August 1995).

BIBLIOGRAPHY

Bureau of Citizens and Cultural Affairs (BCCA) (1990) *The Third Long-term Plan for the Tokyo Metropolis*, Tokyo: Tokyo Metropolitan Government.
Bureau of City Planning (BCP), Tokyo Metropolitan Government (1990) *Housing Master Plan for the Tokyo Waterfront Subcentre*, Tokyo: TMG (in Japanese).
—— (1991) *Guidelines for Townscape in Tokyo Waterfront Subcentre*, Tokyo: TMG (in Japanese).
Bureau of Environmental Protection (1991) 'Report on wide area environmental estimation in 1991', Tokyo: TMG (unpublished) (in Japanese).
Bureau of the Port and Harbour (BPH), Tokyo Metropolitan Governmant (1990) *Data: Countermeasures to the Quicksand Phenomena at the Reclaimed Lands in Tokyo Port*, Tokyo: Tokyo Metropolitan Government (in Japanese).
Government of Japan (1987) *Fourth National Comprehensive Development Plan*, Tokyo: National Land Agency, Government of Japan (in Japanese).
Hayakawa, K. (1987) *Prescription for Living*, Tokyo: Jouhou Shuppankyoku (in Japanese).
Ishii, Y. and Konno, S. (1977) *Coastal Area Development Plan*, Tokyo: Gihoudou Shuppan (in Japanese).
Ishizawa, T. (1987) *Rebirth of Waterfront Area*, Tokyo: Toyokeizai Shinpousha (in Japanese).
Japan Times (1995) 'Aoshima cancels Expo', *Japan Times* (1 June): 1–2.
Koutou Ward, Tokyo (1990) *Basic Survey Report on the Waterfront Subcentre Development*, Tokyo: Toyo Keizai Shinpousha (in Japanese).
Misawa, Y. (1991) *Tokyo Waterfront Development*, Tokyo: Jijitai Kenkyusha (in Japanese).
Moroboshi, K. (1992) 'Kasai Seaside Park', *City Planning* 174: Tokyo: The City Planning Institute of Japan (in Japanese).
Office of Policy Planning, Tokyo Metropolitan Government (1988a) *Basic Plan for the Waterfront Subcentre and Basic Development Policy*, Tokyo: Tokyo Metropolitan Government (in Japanese).
—— (1988b) *Basic Plan for the Waterfront Subcentre and Basic Development Policy for Toyozu and Harumi*, Tokyo: Tokyo Metropolitan Government (in Japanese).
—— (1989) *Implementation Plan for the Waterfront Subcentre Development*, Tokyo: Tokyo Metropolitan Government (in Japanese).
—— (1991) *Toward a Balanced Growth for Tokyo*, Tokyo: Tokyo Metropolitan Government (in Japanese).
Office of Tokyo Frontier Promotion (1990) *Tokyo Teleport Town*, Tokyo: TGM.
Ohtani, S. (1988) *What is Land for Cities*, Tokyo: Chikuma Shobou.
—— (1991) 'Have the courage to turn back', *Architectural Journal* 801: 10–12 (interview in Japanese).
Oshima, T. (1991) 'Have the courage to turn back', *Architectural Journal* 801: Nagoya:

Architectural Journal-sha (in Japanese).
Secretariat of Tokyo Metropolitan Assembly (1987) *The Tokyo Port*, Tokyo: Tokyo Metropolitan Government (in Japanese).
Tajiri, M. (1988) *Proposal: Conservation and Resuscitation of Tokyo Bay*, Tokyo: Nihon Hyoronsha (in Japanese).
Tokyo Metropolitan Government (TMG) (1988a) *Harbour Plan for Tokyo Port, Data no. 1*, Tokyo: Tokyo Metropolitan Government (in Japanese).
—— (1988b) *Fifth Tokyo Port Plan*, Tokyo: Tokyo Metropolitan Government (in Japanese).
—— (1989) *Implementation Plan for the Waterfront Subcentre Development*, Tokyo: Tokyo Metropolitan Government (in Japanese).
—— (1991a) *Guideline for Townscape in the Tokyo Waterfront Subcentre*, Tokyo: Tokyo Metropolitan Government (in Japanese).
—— (1991b) *Outline of the Draft of the Environmental Assessment on Main Roads and Land Readjustment Project in Tokyo Waterfront Area*, Tokyo: Tokyo Metropolitan Government (in Japanese).
—— (1991c) *Report of the Review Panel for the Tokyo Waterfront Subcentre Development*, Tokyo: Tokyo Metropolitan Government (in Japanese).

9

TORONTO: THE URBAN WATERFRONT AS A TERRAIN OF AVAILABILITY

Ken Greenberg

INTRODUCTION

The urban waterfront in the latter part of the twentieth century has become a 'terrain of availability', a screen on which cities project and explore emerging trends and prescriptions for urban development. While 'scripts' for development emerge locally as cities respond dialectically to perceived needs and recent successes and failures, the guiding images projected on to the waterfront have global origins. These reflect the accelerating dissemination of information through journals and professional exchanges and the internationalization of development activity.

In this chapter, a series of major waterfront projects in Toronto will be examined with particular reference to the ideas that led to their generation, the theories that underlie and support them, and their social and political dimensions. Spanning the three decades from the 1960s to the 1990s, seven examples will be considered for the areas known as Harbour Square, Metro Centre, Ontario Place, the St Lawrence Neighbourhood, the Railway Lands (Metro Centre revisited), Harbourfront and Garrison Common (Fig. 9.1). Some of these projects, particularly Harbourfront and Garrison Common, will be discussed from another perspective by Michael Goldrick and Roy Merrens in Chapter ten.

All but one of the waterfront projects are on former industrial land. However, they represent a broad range of approaches to planning and development in terms of their origins, objectives and character. Not all are at the present water's edge (due to an enormous landfill operation carried out in the early part of the century), but most are on land where obsolescence has created a significant and strategic opportunity for urban growth and changes in land use close to Toronto's core.

Before looking at the seven developments, it is worth considering what they represent for Toronto as a whole.

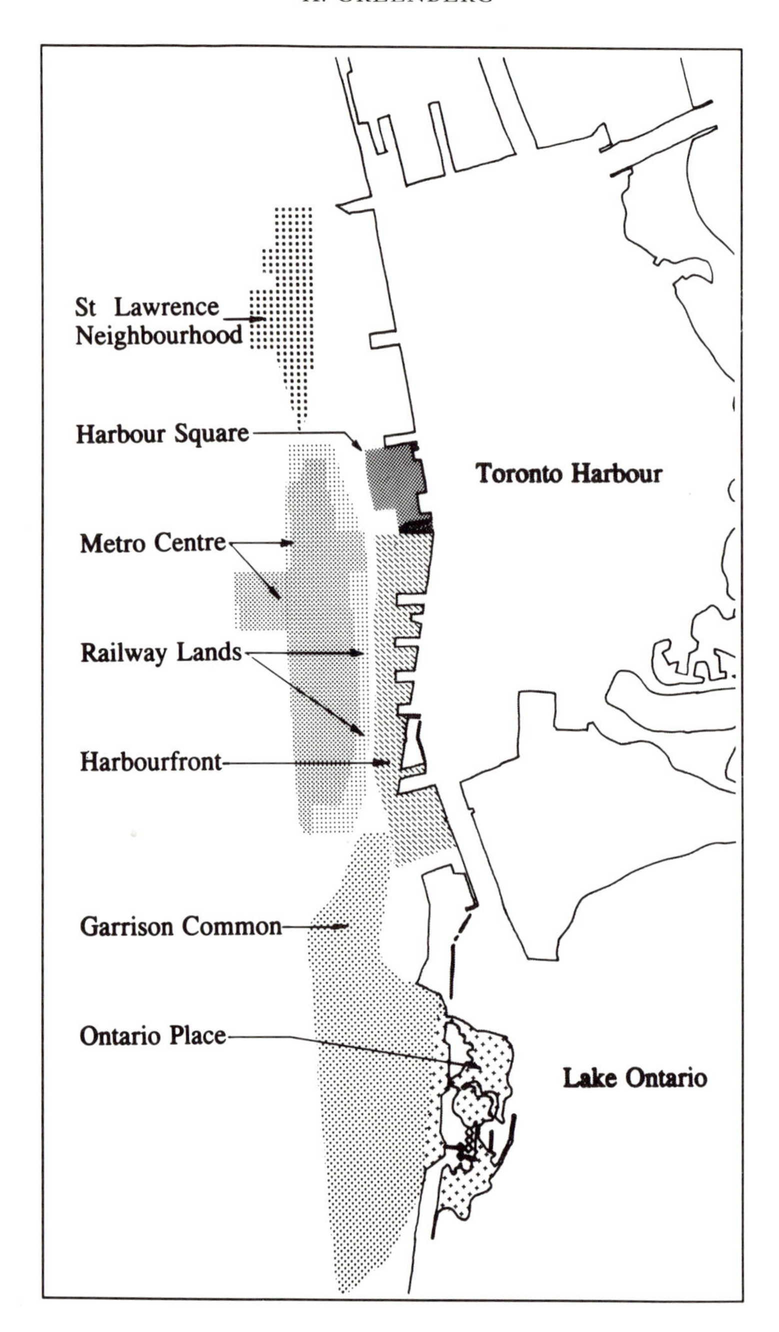

Figure 9.1 Toronto: key plan of the waterfront
Source: Berridge Lewinberg Greenberg Dark Gabor

BACKGROUND

Toronto began as a waterfront settlement on Lake Ontario at the end of the eighteenth century. Its early inhabitants maintained an intimate and vital link with the harbour until the mid-nineteenth century, when railway tracks, laid across the front of the city on low land adjacent to the port, forced out other uses. A project to create a public 'esplanade' on the waterfront was abandoned in the 1840s. Instead, the railway lines expanded to serve a major industrial corridor, which was gradually widened by new piers built out into the bay and by extensive landfilling.

For almost a hundred years the relationship between the city centre and Toronto Bay was interrupted by a port/industrial district reinforced by the opening of the St Lawrence Seaway. There was little reason for the general public to enter the waterfront area, except to use certain services such as the public ferry docks which provided access to the Toronto Islands, a popular place for recreation. Moreover, this relative isolation was promoted by planning policies which sought to segregate urban functions in order to reduce conflict and interference; and which sanctioned, in particular, the separation of work-places from living and recreational areas.

As the city's relationship to Lake Ontario changed, recreational activities moved to peripheral locations east and west of the historic centre, south to the Toronto Islands, and more significantly, into 'cottage country' on lakes several hours' drive from Toronto. Although this centrifugal pattern appeared to be well established, by the 1960s a series of transformations were already under way that would bring it into question. Containerization and the enlargement of ocean-going ships (which could no longer pass through the locks) deprived the St Lawrence Seaway and the port of Toronto of their traditional roles. Changes in rail technology rendered the vast central marshalling yards obsolete. The increased use of road transport and the greater accessibility of suburban and peripheral sites for industries no longer dependent on shipping or rail further destabilized the port. Finally, the globalization of trade and the relative decline of Toronto's industrial base meant that certain activities were relocated to other parts of the world. These factors led to a marked retreat of the port/industrial 'glacier' that had spread across the front of the city.

As the viability of the port/industrial base faltered, the industrial landscape began to develop the now familiar weeds, cracks and rust. Moreover, as land was made available for redevelopment the public perception of the area began to change. The previously unquestioned hegemony of the industrial sector over the waterfront was challenged. A philosophical and political confrontation was brewing which would challenge the prevailing planning ethos based on the radical separation of land uses, buttressed by the building of new urban expressways. This confrontation spilled over to the waterfront, which was increasingly seen as a vacuum and, in turn, as a new frontier for urban development.

To some extent the people of Toronto began to reclaim the waterfront and to demand that it be opened up. They explored this vast *terra incognita* and discovered its charms, claiming nooks and crannies for fishing, makeshift boating clubs and allotment gardens. The view that the water's edge was a valuable public resource to be reclaimed was reinforced by a number of complementary phenomena. A powerful environmental consciousness was developing regarding the extent of damage to soils, air and water caused by decades of industrial pollution and sewage. Whereas bathing beaches had been closed across the entire city, the potential value of the waterfront was increasingly enhanced by the needs of the rapidly expanding population of Metropolitan Toronto, and the growing number of people who did not have access to traditional cottages for summer recreation. Moreover, the lakes within driving distance were becoming over-populated, and the driving conditions for the weekend exodus were intolerable. Thus, the pressure on urban recreational sites was mounting.

Toronto was not alone in experiencing the decline and restructuring of an industrial waterfront. Architects and city planners were particularly interested in historic relationships between cities and water that offered positive models for redevelopment, and by the apparent success of recent waterfront revitalization projects in cities such as Boston and Baltimore.

The combination of industrial decline in the port, and the growing perception of the waterfront as a resource, created the conditions, the market and the opportunity for the series of projects discussed below. These were initiated by the private sector, the government of Canada, the province of Ontario, and the City of Toronto, whether acting alone or in various combinations. Viewed together, the projects can be seen as an evolving sequence of exploratory probes: attempts to find appropriate models for a new relationship between the city and the waterfront as a terrain of availability.

THE SEVEN PROJECTS

Harbour Square (1963–69)

The *Plan for Downtown Toronto*, published by the City of Toronto Planning Board in 1962, drew heavily on proposals which were set out in the report *The Core of the Central Waterfront* (1963) for proposals for the area known as the Lower Edge. To counteract the barriers created by the rail corridor and the elevated Gardiner Expressway that separated the city centre from the lakefront, this plan proposed a series of major interconnecting rooftop plazas leading from Front Street (the historic water's edge) to an elevated pedestrian deck over the tracks south of Union Station. The plan envisaged that, stepping down to the waterfront, the towers rising from these decks would accommodate offices, hotels and other commercial uses. One of the plan's most significant features was that it proposed to avoid difficult ground conditions and create new 'ground' at an upper level.

In response to this invitation to establish a new southern development axis and a 'beach-head' on the waterfront, the Toronto Harbour Commissioners sought private-sector proposals for their land on the edge of the harbour between York Street and Bay Street. After a number of failed attempts during which this land changed hands, the Campeau Corporation, a residential developer, put forward a dense high-rise scheme (Harbour Square, 1969) for the land. A number of residential towers and a hotel were constructed according to this plan (Fig. 9.2).

The Harbour Square proposal was implemented through a complex tripartite agreement between the port authority, the Toronto Harbour

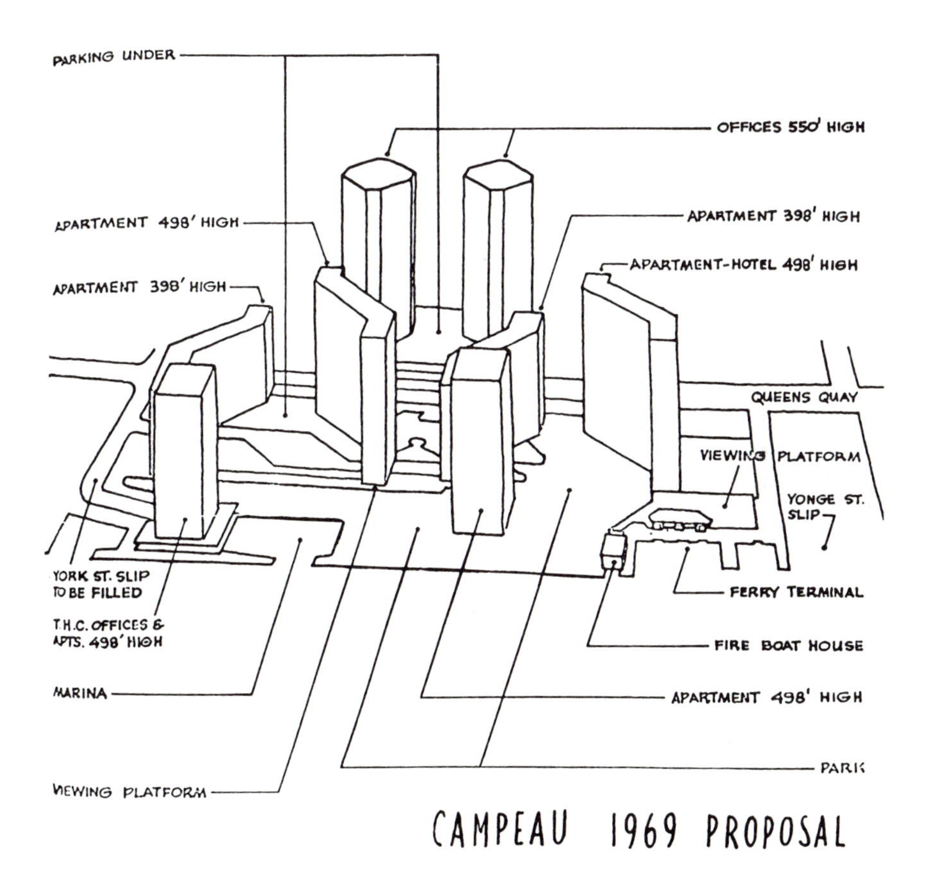

Figure 9.2 Toronto: Harbour Square
Source: City of Toronto

Commissioners (the vendors), the Campeau Corporation (the developer) and the City of Toronto (the planning authority). Faced with declining port activity and revenues, the Harbour Commissioners' objective was clear: they wished to exploit fully what was beginning to be perceived as a valuable land resource, while protecting their interests in the marine operations of the harbour. The Campeau Corporation saw an opportunity to reach a promising new market for luxury condominium apartments and hotel suites with spectacular waterfront views. A pro-development city government favoured this initiative and sought a legal framework that would permit a major change in land use within the port area while protecting essential public operations such as the Island Ferry service.

The initial Harbour Square Plan reflects the clear intention to create a private enclave orientated to the water. Using massive above-grade parking structures as a podium base, a new artificial landscaped ground level was created over the adjacent city streets. This meant that the complex was orientated away from the city's principal waterfront street, Queen's Quay. After crossing Queen's Quay, Bay Street was destined to terminate unceremoniously by plunging underground into a parking garage. Behind the condominium towers there is a 7 acre (2.82 ha) public park on the waterside. This was provided with the Harbour Square project, but it is relatively hidden and inaccessible. The public ferry terminal was demolished, to reappear as an uncelebrated component of the podium base. A hotel, apartment towers and a small convention centre are linked internally by glazed passageways above ground level, making the whole complex functionally integrated, self-contained and separated from the streets below. The towering condominium slabs are angled to maximize views from individual apartments, but they block views from the city to the lake.

In retrospect, the initial 1960s Harbour Square scheme can be seen as typical of a particular approach to urban development. The large building complex was perceived as 'the city within the city'. Its functions were internalized. It evoked vociferous and critical reactions which focused on the massive size of the development, the privatization of the water's edge, the barrier effect of the podium and towers, the lack of public life at ground level, and its microclimate of deep shadows and high winds.

These reactions were to be particularly instrumental in shaping a radically different approach to the redevelopment of an adjacent area known as Harbourfront. Even in the later phases of the Harbour Square development there was an attempt to adjust the 'model' to reflect changing attitudes and priorities. Although the newer buildings were constructed on the same massive scale (reflecting apparently irrevocable development rights) there has been some effort to establish a positive relationship between those buildings and the surrounding streets.

Metro Centre (1968–74)

In May 1968 Canadian Pacific and Canadian National Railways, through a joint-venture company, Metro Centre Limited, released a major planning study for what was then the largest single redevelopment scheme in North America (Fig. 9.3).

The Metro Centre was launched as a major real-estate operation that would involve the total restructuring of transport facilities on roughly 200 acres (81 ha) of railway land. It was to allow the railway companies, such as the Toronto Harbour Commissioners, to exploit the enormous strategic land resources under their control. Unlike the Harbour Commissioners, however, the railway companies proposed to operate through their own real estate companies. The project was also more ambitious than Harbour Square. The Metro Centre Plan set out to alter the entire environment rather than to build a protected enclave. A new consolidated rail corridor was to be created at the southern edge and the city's historic Union Station demolished to make way for major office towers adjacent to the financial core. Union Station would be replaced by a new transportation facility, including a rail and bus terminal integrated with the subway. The world's tallest free-standing tower was to be constructed in the middle of the area. As it turned out, a simplified version of this structure, the CN Tower, was the only part of the Metro Centre scheme to be realized.

The most distinctive aspect of the Metro Centre proposal was the scale of its attempt to integrate regional and urban systems of movement into the city fabric. This was the apogee of the technological city of the 1970s.

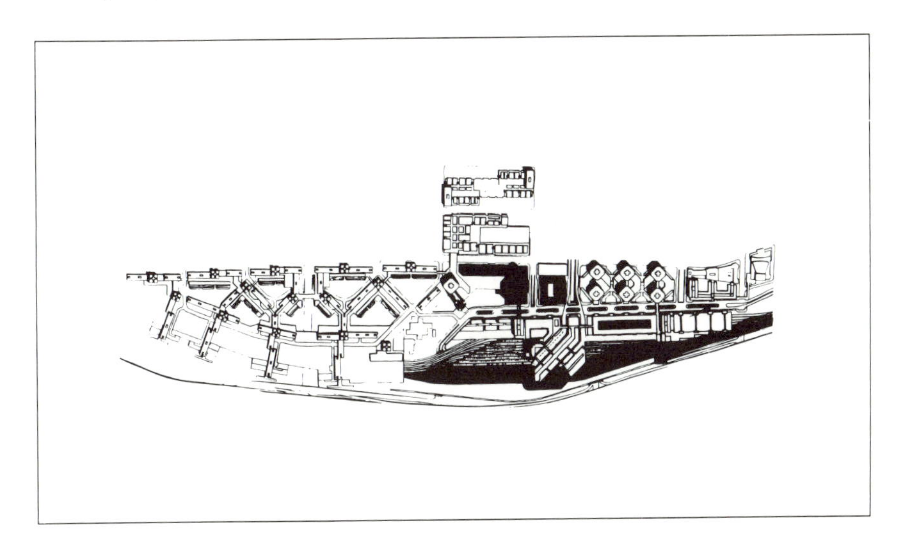

Figure 9.3 Toronto: Metro Centre
Source: City of Toronto

The project was structured so that buildings constituted linked destinations along arteries, while vehicular and pedestrian spaces were interwoven within buildings. The traditional distinctions between building and street were rendered obsolete. The pattern of solid and void in the city was inverted, as buildings embodied streets and the open spaces of the city became residual to buildings. This approach to the urban structure was seen as a means to overcoming the chaos of the modern city by integrating streets, buildings, infrastructures and transportation within 'megastructures'.

The Metro Centre Plan also demonstrated the extensive use of diagonal geometry, a contemporary concept that was intended to facilitate the stringing of buildings on paths or 'desire lines'. While the designers described the Metro Centre as merely a framework or 'organization', in model form it had the powerfully consistent look of Brutalist architecture.

But what made the Metro Centre so extraordinary, and unworkable, was the fact that it tried so hard to integrate a variety of services and transportation systems. The multi-level interweaving of vehicular and pedestrian circulation systems in a diagonal structure departed radically from the north–south, east–west orientation of the surrounding city grid. In so doing, the scheme effectively cut itself off from its context, erecting new barriers between the downtown area and the waterfront.

The Achilles' heel of the Metro Centre proposal, however, was the financial infeasibility of its *sine qua non*, the relocation of the rail corridors. While the plan was approved by the City of Toronto and upheld at a provincial review, debate reopened on its implementation and particularly on the proposed demolition of Union Station. A compromise might have been reached, but these questions were symptoms of a much larger financial problem. In 1974, having invested an estimated $6 million in studies, the railway companies shelved the Metro Centre proposal.

Ontario Place (1967–71)

In 1967 a highly successful World's Fair, Expo 67, was held in Montreal, on a series of islands in the St Lawrence River. Designers from across Canada and around the world participated in the creation of a new kind of relaxed, informal and highly imageable environment on the water. This included structures such as Moshe Safdie's Habitat 67 and Buckminster Fuller's geodesic dome, and it left an indelible impression on the public. A few years later, in 1971, the government of Ontario, inspired by this compelling new model of urban progress and sociability and by an impending provincial election, opened Ontario Place on the shores of Lake Ontario (Fig. 9.4).

Whereas Expo 67 was a temporary exposition, Ontario Place was a permanent but seasonal playground, a political showcase in which a bounteous and successful provincial government could display its wares and entertain the public. Combining marinas, theatres, beer gardens and exhibition halls,

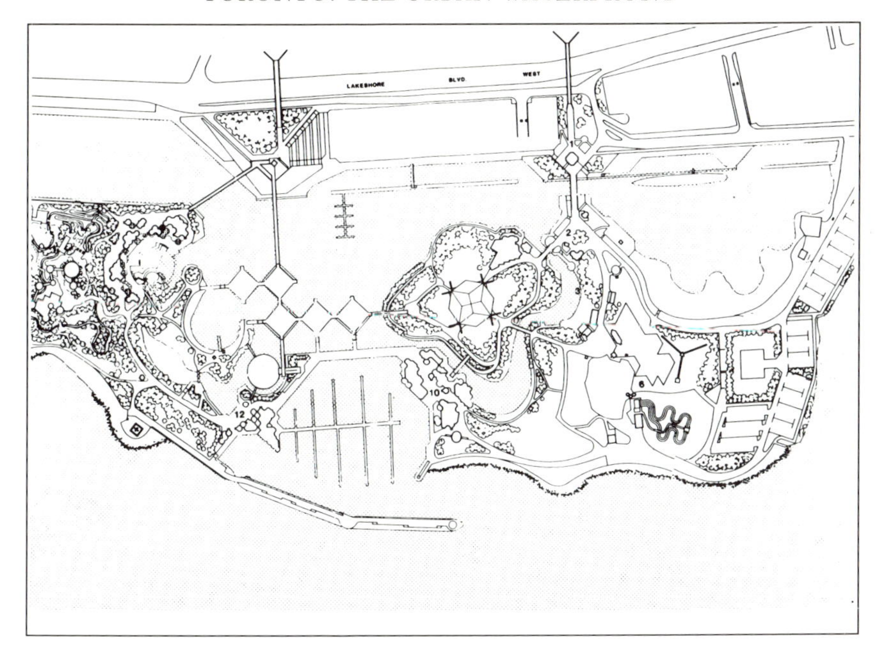

Figure 9.4 Toronto: Ontario Place
Source: Berridge Lewinberg Greenberg Dark Gabor

it was to be light-hearted and fun, but also didactic and market-orientated. The image of technological sophistication and social ease was also intended to suggest a new relationship between the public and Lake Ontario.

Ontario Place was built as a series of man-made islands by employing the newest techniques for landfill. It took the form of a giant cruise ship anchored offshore and adjacent to Exhibition Place, the traditional urban fairground. Ontario Place has ticketed points of entry and a series of pathways linking a sequence of 'pavilions'. The geometrical and modular structures of Ontario Place were strongly reminiscent of Expo 67. There were echoes too of Expo in its large exhibition 'pods' on stilts over the water, its small-scale boutiques and eating areas, and its supergraphics and landscape of curvilinear moulded hillocks and mass plantings. Its offshore location also provided the opportunity for a highly visible skyline reading as a kind of futuristic and technologically precocious 'Magic Kingdom' set in a garden by a lake.

The theme of this 'kingdom' was Ontario, its people, resources, affability, wealth and potential for growth. This was not an élite environment, but neither was it free. Popular, and aimed at the broad middle class, Ontario Place was orchestrated by a provincial corporation to open a window on the province and to reflect a new spirit and sense of possibility.

Although initially highly successful, Ontario Place has not aged well, perhaps because of the attempt to give permanence to the inherently ephemeral. Parts of the development, such as the popular 'Forum' – an outdoor stage for summer entertainment – and the Children's Village, have enjoyed continued success. As a whole, however, Ontario Place now seems a somewhat dated attraction, lacking the allure of novelty or of a uniquely privileged position on the water's edge. Its seasonal character and isolation have begun to appear problematic, wasteful and limited.

There is now a wish to redefine the role and form of Ontario Place. Under consideration are: integration with its larger urban context; probable year-round access; a continuous waterfront trail for pedestrians and cyclists; more access by public transit; and a removal of the physical barriers that isolate it from the adjacent Exhibition Place. These issues have been examined recently in a plan for the larger area known as Garrison Common.

St Lawrence Neighbourhood (1972–74)

In a series of municipal elections of the late 1960s, and through a significant victory in 1972, a 'reform council' took over at Toronto City Hall. Confronted by the planning and development practices of the preceding period, it was committed to the defence of neighbourhoods and to halting the expansion of urban expressways, while promoting public transit and pedestrian environments. One of the first tasks of the reform council was a review of the city's plan for the downtown core, a rapidly expanding central business district of high-rise office towers surrounded by parking lots.

The resulting new *Central Area Plan* (Toronto Planning Board 1974) proposed to curtail office growth in downtown Toronto, by distributing space to emerging suburban centres. It also aimed to match office growth to the capacity of the public transit system, and to introduce provisions for mixed use that would bring a substantial new residential population into the core.

These ideas were fiercely contested by a development industry which was used to having its own way. In particular, there was considerable scepticism about the possibility of attracting people to live in a downtown environment. The burden of proof fell upon the City Council. Led by Mayor David Crombie, a strategic demonstration project was conceived for a block of obsolescent industrial land on the historic waterfront to the south-east of the financial core. Forty-four acres (17.74 ha) were expropriated and plans prepared for the St Lawrence Neighbourhood, which would eventually provide housing for a diverse population of 10,000 within walking distance of the core (Fig. 9.5).

From the outset, the inspiration for this project was a contemporary reinterpretation of the traditional Toronto neighbourhood. Given a dense and urban form, this was seen as an answer to the failed modernist vision of towers in the park. Inhabitants were to be of mixed income to break the

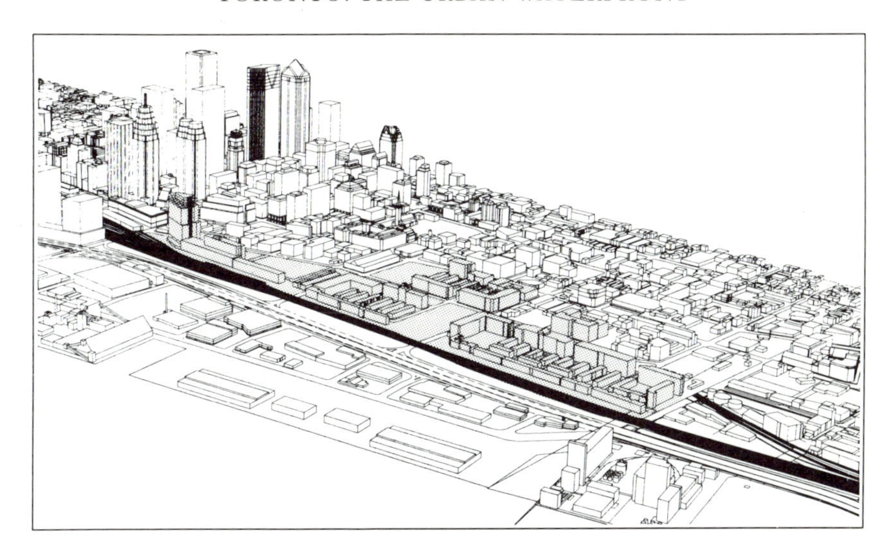

Figure 9.5 Toronto: St Lawrence Neighbourhood
Source: City of Toronto

mould of previously segregated public housing projects. An unused rail corridor that traversed the site became the leafy linear Crombie Park and promenade leading to the St Lawrence Market and the core. Leading off this central spine were a series of rectangular blocks with rows of parallel apartment buildings lining the park; behind these blocks, townhouses were inserted in a form intended to echo the historic Toronto pattern.

While the design *parti* had strong local roots and neighbourhood precedents, it also exhibited clear affinities with a series of contemporary European projects that attempt to organize new housing districts as urban 'quarters' on former industrial or other vacant sites (such as those in Berlin, Paris and Barcelona). Many non-profit-making and co-operative housing developers and architects were involved in the St Lawrence project. Architectural quality was an issue, particularly in the early phases, and urban design criteria were strict. This project may represent the closest Toronto has ever come to achieving design coherence in a large-scale, phased renewal project.

In most respects the St Lawrence Neighbourhood has been hugely successful. Not only did it succeed in creating an environment which is much admired and enjoyed by residents and visitors, but it spawned a series of private-sector initiatives on its borders, providing additional market housing, and infused significant new life into an area that had been largely uninhabited for about a century. The development of the St Lawrence Neighbourhood also showed the potential for a new residential population in the core. It revived the St Lawrence Market, and set the stage for a continuing process of revitalization in adjacent districts. Most significantly perhaps, it demonstrated the fact that a mixed-income population could live together in harmony.

If there is a criticism to be levelled at the St Lawrence Neighbourhood, it is probably that it is almost exclusively devoted to housing. There are few shops and little or no employment or other activities, but these defects could be remedied when the remaining peripheral sites are redeveloped.

It is curious that the positive lessons from the St Lawrence Neighbourhood experience seem to have had a limited impact in terms of other development. The particular circumstances that gave rise to the St Lawrence Neighbourhood would be difficult to duplicate, namely the combination of strong political leadership, a dedicated and knowledgeable planning team and the backing of social housing programmes and federal mortgage financing. One of the chief factors, however, in the success of this new neighbourhood was the willingness to define a clear plan, to entrust leadership for its implementation to a responsible project group and to allow that group to carry out its mandate over a considerable period of time. This ability to conceive an approach and let it go the distance has proved highly elusive in subsequent attempts to redevelop the waterfront – as witnessed by the plans for the Railway Lands.

Railway Lands development concept (1983)

After the demise of the Metro Centre, it was not surprising that the next phase of planning for the Railway Lands involved a loosening of the Gordian knot that had bound land-use and design concerns to ambitions for the improvement of regional transportation. The province of Ontario, facing enormous transportation pressures, had already set in motion an independent programme of works to accommodate the emerging GO commuter railway system, based on the retention of Union Station and the northern rail corridor. The province also established a Land Use Committee with representatives from all levels of government and the railways to rethink the land-use plan in this new context.

The political environment had also changed in the interim. The legacy of the reform movement was a determination not to acquiesce in privately initiated development plans, but to reshape them in terms of the public agenda set out in the new *Central Area Plan*. Attempting to reassert its control over the next round of planning for the Railway Lands, the City of Toronto adopted a document called the *The Railway Lands: Goals and Objectives* (Toronto Planning Board 1978). Among a host of specific objectives it stated that the Railway Lands should be used to reconnect the city to the waterfront, that there should be a broad range of uses, and that development should be street-orientated and based on traditional urban patterns.

In September 1983, staff in the city's Department of Planning and Development, after extensive consultation with railway companies and other agencies, put forward the city's own concept which was adopted by Council as the basis for planning and an implementation strategy (Fig. 9.6). This concept represented a clear attempt to strike a new balance between property and public interests. What distinguished it from the Metro Centre in design terms was the reintroduction of the concept of the traditional street as the basic structuring device. The street was re-established as a traditional device for vehicular traffic, public transit systems, pedestrians, services and utilities; and also as the means of subdividing land into building sites. Without the burden of overlapping multi-level traffic routes, this approach also permits the construction of traditional building types.

This shift was not solely a site-specific reaction to the problems encountered in implementing Metro Centre. It reflected a broad and dramatically different consensus on the role of streets, first articulated in the city's *Central Area Plan*, and now evident in attitudes towards redevelopment in all parts of the city. The Railway Lands were unusual, however, in that the area had few streets and, therefore, new streets had to be introduced from scratch.

A second important change was the reintroduction to the urban repertoire of the traditional urban park and square and public open spaces defined by building facades. As opposed to the irregular and often formally and functionally ambiguous semi-public green spaces left over at the base of free-standing towers, the public parks in the Railway Lands Development Concept were clearly intended to be publicly maintained urban spaces.

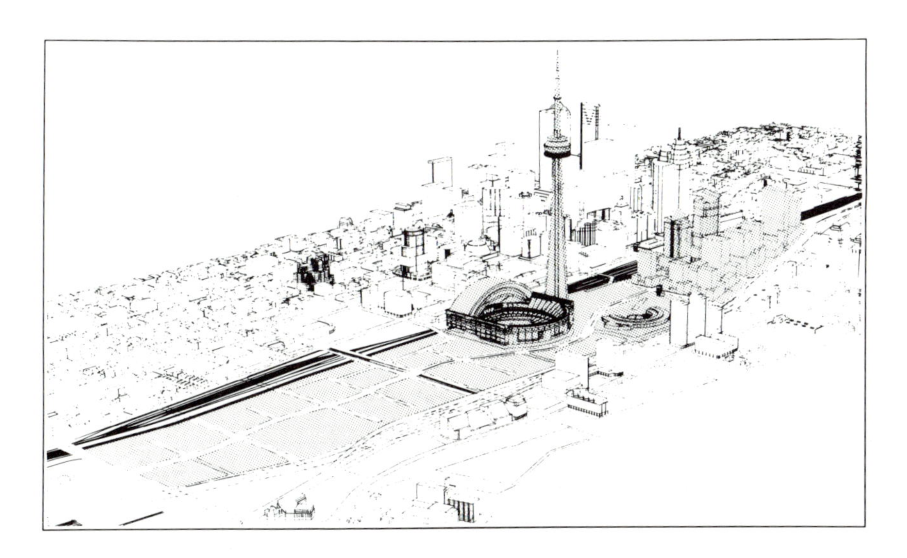

Figure 9.6 Toronto: Railway Lands
Source: City of Toronto

Another new issue peculiar to the Railway Lands Development Concept was the recycling and reuse of certain elements of railway heritage. Whereas the Metro Centre plan had proposed to make a *tabula rasa* by eliminating the artefacts of the railway era, including Union Station, the Railway Lands Development Concept embodied a new sensibility. Based on the fact that it was not possible to reorganize everything at once, and that it was beneficial to retain some ties with the past, the concept plan was organized so that some of the most significant patterns and structures built by the railways could survive.

In a sense, planning had come full circle with the plan for the Railway Lands. The first response (Metro Centre) to developing this area was to make it as different as possible from the existing city. The revised concept plan suggested that it be modelled on the successful aspects of Toronto's traditional urban form.

In the public discussion of the Railway Lands Plan, the major issues raised were concerned with scale and density. Would the amount of development proposed for this area result in buildings that were too large? Would it have a negative impact on plans for suburban centres? Would it necessitate the construction of major new regional road and transit facilities? Would there be enough housing, or too much commercial development?

With changes in the political administration, and after a prolonged and deep recession, previous agreements about amounts and types of development were renegotiated. Contrary to previous expectations regarding an extension of the financial district, it now appears that new housing and entertainment will be the predominant uses. This testifies to the virtue of a street and block plan that can adapt to proposed changes in land use.

The only significant structure yet built on the Railway Lands is Skydome, a stadium with a retractable roof, which has been developed together with adjacent streets and blocks. The decision to locate this stadium close to the heart of the financial district and the hub of the regional public transport network (and to provide it with minimal parking) was controversial, but the stadium appears to work well in practice, bringing added life to the core after office hours.

Harbourfront (1972–76)

Of all the areas considered here, Harbourfront is the most complex in terms of its multiple and changing agendas and high-profile successes and failures (Fig. 9.7). Introduced in 1972 as a federal gift to the City of Toronto, Harbourfront was presented as a relatively simple idea: an 86-acre (35 ha) park to replace moribund industries on expropriated land along a stretch of the harbour from York Street to Bathurst Street. Similar federal

Figure 9.7 Toronto: Harbourfront
Source: City of Toronto

initiatives were launched at approximately the same time in Vancouver, Montreal, Halifax and Quebec City – each has had a complex history.

Toronto's Harbourfront 'park' was first seen by the federal authorities as an antidote to the over-development and sealing off of the waterfront that had occurred next door in Harbour Square. On closer inspection by the city, however, it was thought that a conventional 'grass and tree' park would be too exposed and isolated and, if it was to provide a broad range of facilities, too expensive to maintain. During long public consultation the idea of a park was gradually replaced by a more complex concept that included a 'host community' with a residential and working population – similar to earlier examples of Toronto waterfront communities. Other objectives, such as capitalizing on the potential of some existing industrial structures, also gained momentum. Meanwhile, an extraordinarily successful arts and culture programme began to take root in the vacant indoor and outdoor spaces of Harbourfront.

Pressure from a new Conservative government in Ottawa, and initial experimentation in a very strong real-estate market, led to the idea that development revenues could offset site improvement costs *and* pay for cultural programmes. Eventually, a Crown Corporation was set up, with a local board to act as site developer. A development framework was produced to guide private-sector involvement. This framework for Harbourfront institutionalized the following progression: from the concept of a park, to that of a park with some buildings, to an intensive waterfront development with an above-average amount of parkland, including a water's-edge promenade and an unusual concentration of cultural facilities.

There were some notable early successes in blending public and private objectives and resources. A major renovation, carried out by Olympia and York, transformed a 119,000 yd^2 (100,000 m^2) warehouse (the Terminal Warehouse) into Queen's Quay Terminal: a mix of apartments, offices, shopping, a contemporary dance theatre and a lively water's edge. A former trucking terminal became a dynamic cultural hub with a myriad of successful programmes and events, including aspects of all art forms from popular to high culture, from children's festivals and day camps to a world-renowned authors' festival. However, as the economy faltered, sometimes there were tensions between the competing priorities and conflicting expectations of Harbourfront – as an aggressive development project and as a popular cultural venue. These tensions were severely exacerbated by the recession of the early 1980s, when the goal of financial self-sufficiency became unattainable. By this time it was no longer clear what the true objectives of the Harbourfront exercise really were. With no commitment to public funding of the programmes, the only source of funds was more and more development. Expedient and highly unfortunate development deals were made, resulting in a series of extremely unpopular building projects, which undermined public confidence in, and support for, Harbourfront's management.

Work on public spaces lagged behind due to the decline in financial resources, and when the projects were completed the results were sometimes disappointing. For example, conflicts arose between Harbourfront, the City of Toronto and Metropolitan Toronto, over the role and character of Queen's Quay Boulevard. This should have been Harbourfront's premier public space, but it was developed as an excessively wide, over-designed automobile and transit artery with little amenity value or appeal. In the end the crisis was such that the Harbourfront organization was disbanded when the project was half complete.

An attempt has been made to salvage some of the cultural programmes associated with Harbourfront, albeit at a much reduced level. The remaining building sites have been parcelled out to the private sector in a series of real-estate deals. There has been little attempt to update the urban-design strategy, although development is still lagging behind because of the state of the economy. Whereas this may change soon, the sole remaining planning objective seems to be to maximize the open (unbuilt) space on the water's edge and to exploit the development potential of a smaller number of sites.

In its unfinished state, Harbourfront is a puzzling combination of brave, ambitious and unfulfilled promises. The project once suggested a highly innovative way of crossing the boundaries between a commercial real-estate operation, and a non-profit-making cultural organization. But these were flawed in their execution and, under pressure, proved to be unworkable. Harbourfront does combine entertainment, living and working, although in an embryonic form, but is still extremely isolated from the rest of the city. There is a streetcar, but walking connections are difficult or impossible. Parking lots have been zealously exploited for profit, while a bicycle path, which could have been an important feature and means of access along the waterfront, was omitted from the plan. Harbourfront's architecture combines the good, the bad and the indifferent; its urban design, so far, has no discernible overall image, except the confusion that results from numerous shifts in direction.

In short, Harbourfront is a poignant partial success. It exhibits all the conflicts and contradictions of its complex and arrested origin, including the inability to deal with an economic recession, the failure – so far – to resolve conflicts between public and private interests and between agencies and levels of government, and the failure to prioritize objectives and create a plan with staying power.

Garrison Common (1991)

Garrison Common, the name given to a small nineteenth-century common, is now used for a much larger district of Toronto's downtown core (Fig. 9.8). It includes some of the city's most significant cultural sites: Fort York, Exhibition Place, Ontario Place and Coronation Park. This district provides a major facade for the waterfront, but individual sites (including significant areas of under-utilized land) are poorly connected and enmeshed in a web of prohibitive transportation barriers. The Garrison Common Master Plan was prepared in 1991 for the Royal Commission on the Future of the Toronto Waterfront (as an independent and unbiased agency, theoretically operating without vested interests) to address the possibilities that could be generated by pooling land and reintegrating planning over this larger area.

It was widely recognized that discrete projects could no longer work effectively. Problems and potential solutions transcend ownerships and jurisdictions. This is evidenced by the fact that the historic Fort York, a site of exceptional interest and beauty, is virtually inaccessible in a 'trench' that is bounded by an elevated expressway and the rail corridor. Ontario Place and Exhibition Place are divided by Lakeshore Boulevard, a busy arterial road with security fences. Similarly, hectares of marginal parking lots and obsolescent industrial sites express a decline in traditional uses and the need for a new collective vision.

The Master Plan adopts a larger perspective starting from a re-examination and redefinition of the international, regional and local roles of the Garrison Common area. It established the area's primary roles as cultural and recreational, with additional proposals for the creation of new residential

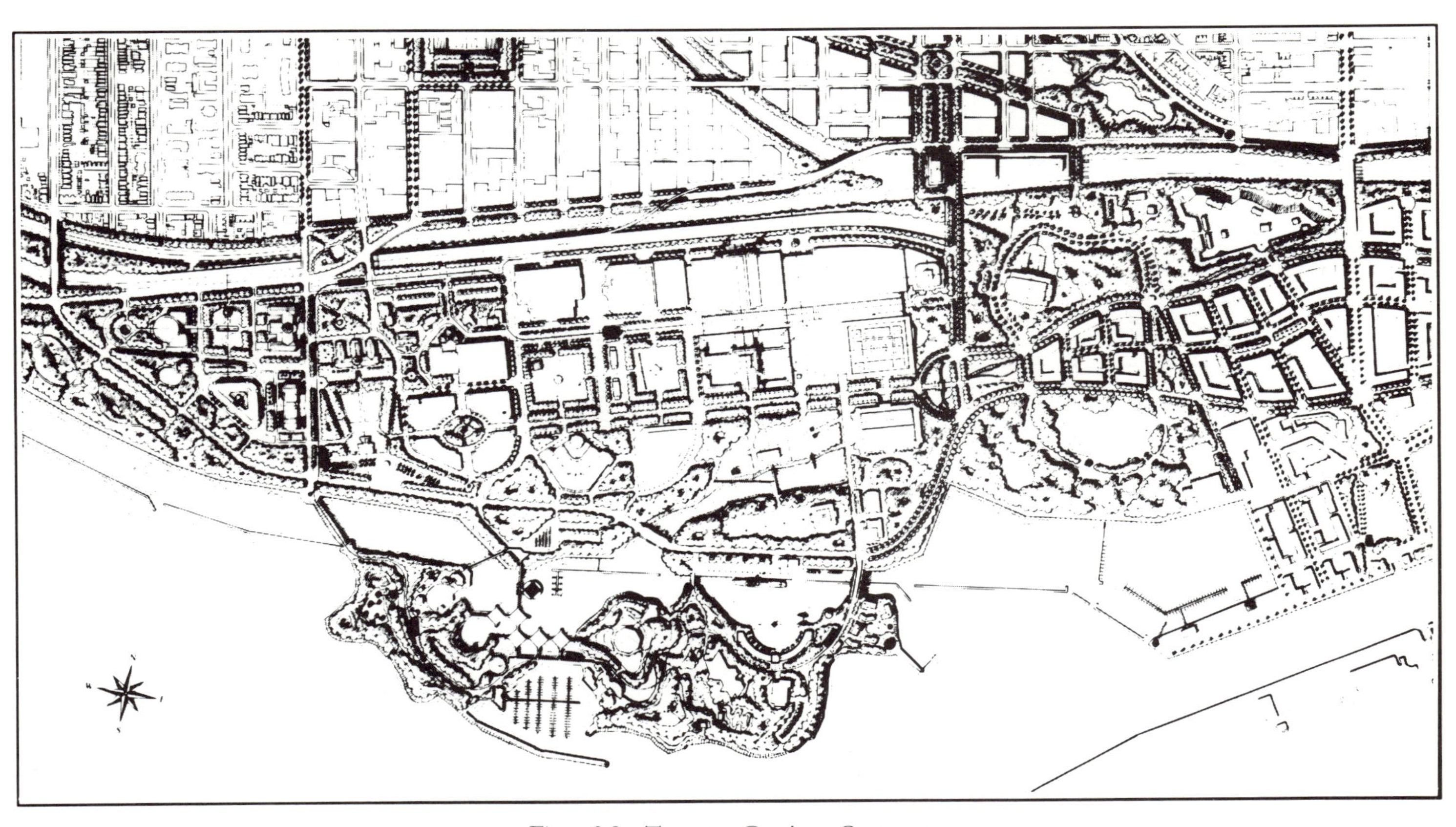

Figure 9.8 Toronto: Garrison Common
Source: Berridge Lewinberg Greenberg Dark Gabor

communities and employment centres. The plan also generates an overall environmental framework of remedial action on the whole site to deal with water, soil and air quality. In addition, it proposes a large-scale reassessment and restructuring of transportation facilities.

Rail traffic was meant to be consolidated in one of two existing parallel rail corridors, freeing up the other corridor to become a new linear park and the spine of a new residential neighbourhood. New roads were to be introduced to permit the scaling-down of existing roads, such as Lakeshore Boulevard. The possibility of partially burying the elevated Gardiner Expressway was also explored. The distinctive landscapes of Ontario Place and Exhibition Place were identified, and new physical linkages and combined planning for events and programmes were proposed. Cross-boundary improvements, providing access to the lake from existing neighbourhoods and along the lakeshore, were also introduced.

The Garrison Common Master Plan departed from the plans examined above in that its primary focus was not on the development of specific areas or facilities but on the issues of connection, linkages and the urban fabric. It cut across property, functional and disciplinary lines. In employing an 'ecosystem' approach, it represented a tremendous challenge to vested interests, including the large private owners of former industrial properties and the agencies of several levels of government. While the need for such an overriding approach was widely acknowledged in theory, and the plan has been received favourably in principle, this approbation has yet to be translated into concrete action. The clear implication of the Garrison Common Master Plan is that new and different media will be required for its implementation. It remains to be seen whether they will emerge. At present, the only significant initiative under way in this area is a major new Trade Centre.

In some respects, the Garrison Common Master Plan is still a plan-in-waiting. It expresses broad intentions but has a less well-defined three-dimensional character and image than the other plans discussed. This lack of definition reflects its larger scale and the recognition that its physical embodiment must await the resolution of major issues and the interplay of powerful forces.

CONCLUSION

The seven projects described above do not represent all of the efforts to revitalize the Toronto waterfront since the late 1960s, but they certainly constitute a significant proportion of them, and are broadly indicative of the ideas and approaches that have informed those efforts.

Through their designs and programmes, the seven projects opted for particular visions that addressed not only specific waterfront issues, but took implicit or explicit positions on the major questions of urban form that were under debate during their respective periods. In so doing, each project embraced one or more principal ideas and assumed a particular character or emphasis. A partial list of these (sometimes overlapping) basic issues is explored below. While sometimes presented in terms of polarities, in fact the positions taken in development projects frequently fall somewhere between pure extremes.

Building blocks: the nature of the urban framework

Waterfront sites are typically large and set in former port and industrial areas that lack basic urban infrastructure, streets, utilities, public transport, community amenities, etc. This raises questions about the nature of the new urban framework implanted into redundant port/industrial areas. It raises issues, for example, relating to the definition of individual building sites and the nature of new streets and public spaces. The different responses to these questions among the seven projects mirror an ongoing theoretical debate. They reflect a variety of positions on urban morphology: from the modernist concept of the superblock, with pedestrian and vehicle circulation at different levels, to the traditionalist approach, with a familiar framework of streets and blocks. Moreover, they represent different views about the extent to which waterfront projects are seen as 'extensions' of the existing city fabric, or as new and dramatically different urban 'ensembles'. This question of compatibility is addressed at the related levels of urban form and architectural expression.

Industrial heritage: liability or opportunity

Undoubtedly, waterfront redevelopment has broadened the range of heritage preservation. Virtually all urban waterfront sites have experienced one or more generations of previous use, frequently as port/industrial sites. In most cases there is a legacy of buildings, structures, landforms and archaeological artefacts. In Toronto, in general a hasty *tabula rasa* approach has been taken to this industrial legacy. Recently, some projects have incorporated and recycled these elements to great advantage. However, there remains a divergence of opinion on whether the few remaining massive industrial structures of this century (such as grain silos) should be respected in principle, perhaps as monuments to a heroic past, or removed where they are not obviously usable or impede the introduction of new uses. At this time, some are being demolished and others passionately defended.

FUNCTIONAL STRUCTURE

The programme: revising the urban land-use mix

Responding to demands for a range of new facilities, these waterfront projects have contributed to the debate on the appropriate scale and grain of combined uses within the urban structure. New hybrid development concepts have emerged from the distillation and combination of a number of uses such as amusement parks, shopping centres and urban neighbourhoods. Of particular interest in this respect is the extent to which waterfront projects are seen either as ordinary places made special by virtue of their waterfront location, or as overtly atypical and distinctive places. In any event, one factor which has emerged clearly is the need for flexibility, given the time that it takes to realize projects and the changing market forces.

The tradition of waterfront promenading, and the presence of visitors, have added impetus to the idea that the arts (in terms of the performing and visual arts, popular celebrations, high culture, entertainment, crafts, etc.) have an important place in contemporary waterfront projects. The value and extent of public elements varies enormously among the seven projects examined above. In some cases projects are aimed at highly commercial tourist-orientated markets. In others they play a vital role in the local cultural scene.

Social equity: whose waterfront?

The issue here is whether the public has a stake in the waterfront as a resource belonging to all. Given that 'privatization' of the waterfront has often been contested, beyond the frequent question of public access lies the issue of which uses should occupy privileged waterfront sites. Should the land-use structure follow simply from market forces, which award sites on the basis of the ability to pay, or should planning mechanisms be used to broaden the constituency of users?

The varied responses to these questions among the seven projects examined above should be viewed against the backdrop of a broader debate regarding the appropriate role of the public and private sectors, competing claims on scarce government resources, and the recent emergence of a neo-conservative political agenda.

Transportation: questions of priority

Waterfront sites have frequently accommodated major transportation corridors which pass through them but which provide poor local accessibility or impede access to the waterfront. Thus, issues arise regarding the nature of linkages, whether to adjacent areas or within the sites. There is the issue of access to and from other parts of the city, and the question of continuous access along the waterfront. The weighting that each project accords to different transportation modes, cars, transit, bicycles, etc., gives a clear indication of these priorities.

Development process: making plans for what and by whom

This issue has several interrelated dimensions. First, because waterfront sites generally fall outside the boundaries of traditional planning districts and raise many technically complex new issues, they beg the question of whether the traditional tools of planning and the assignment of responsibilities are appropriate. Second, there is an inherent tension between the desire to create an efficient special authority empowered to face new challenges and produce results, and the commitment to democratize the planning process, ensuring that all stakeholders have a voice. This is further complicated by the multiplicity of government layers and institutional and private interests. In addition, there is the issue of the appropriate roles of the public and private sectors. Are they partners? If so, who is responsible for what? How are the competing priorities of private investors and society as a whole balanced in the process?

Finally, there is the changing nature of the waterfront plan itself as a visionary and operational tool. How prescriptive and how proactive should it be? What degree of flexibility needs to be preserved for unforeseen contingencies or the vicissitudes of the market-place? The seven projects demonstrate a broad range of possibilities.

The ecosystem: a broad framework for repair and reconciliation

As a more complete understanding of the interrelated and complex effects of human interventions on waterfronts has emerged, there has been a corresponding and necessary increase in caution. Issues arising from the pollution of land, air and water, and the consequences of altering shorelines by introducing landfill have highlighted the shortcomings of a compartmentalized approach to problem solving, and the need for a holistic and integrated planning methodology that transcends traditional disciplines, site boundaries and political jurisdictions.

It might be tempting to see the seven projects as gradually confronting these issues in a chronological progression, improving with each step. But such a reading would be too simplistic. Certainly a 'learning curve' can be discerned with respect to certain issues, such as the shift from triumphant megastructures to more practical public frameworks of streets and open spaces and the increased understanding of environmental impacts. It is also true, however, that the locus of the problem keeps shifting as new and unpredictable challenges arise and the political economy of development focuses on new opportunities. The city is, after all, dynamic and continuously evolving.

While this chapter has focused on issues of urban design in physical terms, these issues are not considered and resolved in isolation. Physical design is just one factor in the larger debate in which economic, political and social interests interact and compete within the 'terrain of availability'. Ultimately, the physical attributes of each project reflect and embody the specific set of ambitions prevailing in that project – a particular mix of profit motive, political will, social demands, ecological ambitions, etc. Projects also reflect broader ideologies and cultural movements such as neo-conservatism, the drive towards privatization, the lure of megaprojects, the urge for social reform and equity, ecological concerns, and so on.

If the waterfront in the late twentieth century is viewed as a 'testing ground' for new ideas about urban form, it is only through the embodiment of various approaches in plans and actual projects that the strengths and weaknesses of those approaches are finally understood and evaluated. As is the case with most experimentation, the failures are often as informative as the outright successes.

BIBLIOGRAPHY

Ashton, B. and Kinahan, F. (1993) *Draft Metropolitan Waterfront Plan, Toronto: Metropolitan Toronto* (Ontario: Regional Municipality) Planning Department.

Berridge Lewinberg Greenberg *et al.* (1991) *Garrison Common: Preliminary Master Plan, Publication no. 14*, Toronto: Royal Commission on the Future of the Waterfront.

Hack, G. and Carr, L. Associates (n.d.) 'Recharting a course for Harbourfront: an assessment of Harbourfront planning and design' (unpublished).

Harbourfront Corporation (1987) *Harbourfront 2000: A Report to the Futures Committee of Harbourfront*, Toronto: Harbourfront Corporation.

Toronto (Ontario) Planning Board (1962) *The Plan for Downtown Toronto*, Toronto: Toronto (Ontario) Planning Board.

—— (1963) *The Core of the Central Waterfront: A Proposal by the City of Toronto Planning Board*, Toronto: Toronto (Ontario) Planning Board.

—— (1970) *Metro Centre*, Toronto: Toronto (Ontario) Planning Board.

—— Toronto (Ontario) Planning Board (1974) *Central Area Plan*, Toronto: Toronto (Ontario) Planning Board.

—— (1978a) *The Railway Lands: Proposed Goals and Objectives*, Toronto: Toronto (Ontario) Planning Board.

—— (1978b) *The Railway Lands: Basis for Planning*, Toronto: Toronto (Ontario) Planning Board.

—— (1983) *The Railway Lands Part 2: Development Concept*, Toronto: Toronto (Ontario) Planning Department.

—— (1985a) *The Railway Lands Part 2: Implementation Strategy*, Toronto: Toronto (Ontario) Planning Department.

—— (1985b) *The Railway Lands Part 2: Memorandum of Conditions*, Toronto: Toronto (Ontario) Planning Department.

—— (1994) *Garrison Common North Part 2: Official Plan Proposals*, Toronto: Toronto (Ontario) Planning Department.

Waterfront Regeneration Trust (1993) *Garrison Common Implementation Plan*, Toronto: Waterfront Regeneration Trust.

10

TORONTO: SEARCHING FOR A NEW ENVIRONMENTAL PLANNING PARADIGM

Michael Goldrick and Roy Merrens

INTRODUCTION

Waterfront redevelopment during the last two or three decades has been plagued by institutional fragmentation and bureaucratic obsolescence. These problems have become increasingly evident as traditional waterfront uses have declined, as patterns of land ownership have been reconstituted and public agencies have become redundant or have been replaced. In a transitional period of waterfront redevelopment, institutions, jurisdictions and processes have been replaced and rethought.

Difficult adjustments associated with these changes have recently been overlaid by attempts to confront the challenge of ecologically sensitive urban restructuring. The paradigm shift that this new concept represents requires that urban structures be redesigned to achieve compatibility between cities and the natural world. Yet the way in which operational frameworks for such a process should be conceptualized, let alone structured, is still largely unknown.

This chapter presents an analysis of the emergence of an innovative approach to waterfront planning and policy-making. It makes particular reference to a Canadian Royal Commission which attempted to grapple with these questions. Initially, the Commission defined the problems associated with waterfront redevelopment in conventional terms, and prescribed traditional administrative solutions to them. But gradually its understanding of these problems became more complex as the Commission embraced the concept of ecosystem-based urban restructuring. The remedies that it then proposed reflect something of the scope and scale of this new planning paradigm.

At its root the ecological challenge confronting cities as they proceed through various stages of 'creative destruction' has less to do with administrative solutions and more to do with moral values and social choices. As Hahn points out, cities are 'built thought' and 'represent the most materialized form of a society's interaction with the natural environment' (Hahn 1991: 7). In the planning of cities, *political* alternatives must be found to the social and economic processes of industrialized society that threaten the integrity of the environment. Conceptually, the Royal Commission placed urban waterfront planning in the novel context of environmental sustainability. But in practical and operational terms it settled for administrative reforms.

BACKGROUND

The first major planning effort on Toronto's waterfront was launched early in the twentieth century. The Toronto Harbour Commission – set up by the federal government in 1911 to replace an ineffective earlier waterfront authority – promoted and implemented an elaborate plan that provided a framework for redevelopment for almost half a century (Merrens 1988). In the process of implementing this plan, the mouth of a major river flowing into the old waterfront was diverted, large marshes and wetlands were destroyed, and extensive areas along the shallow edges of Lake Ontario were filled in. The remodelled waterfront functioned as the setting for an array of shipping and rail facilities, bulk storage sites and industrial activities typical of many other contemporary urban waterfronts, although embellished with recreational activities on the flanks of the working waterfront.

Starting in the late 1950s, Toronto's waterfront became the object of renewed attention. A few modest and informal planning ventures were undertaken in the 1950s. Then, in 1962, a plan entitled *The Core of the Central Waterfront* (City of Toronto, Planning Board 1962) was presented by the City of Toronto. This turned out to be the first of a number of major proposals for reshaping the waterfront that appeared over the next three decades. The latest of these is in the form of the final report of the Royal Commission on the Future of the Toronto Waterfront (hereafter Royal Commission), which emerged in 1992 under the title *Regeneration.*

The lavish attention accorded waterfront planning in Toronto in recent decades is representative of a larger change affecting urban waterfronts in many places. During the second half of the twentieth century, waterfront dereliction and redevelopment have accelerated world-wide as the consequences of economic restructuring have become more pervasive. Like many other waterfront cities, Toronto found itself with extensive areas of abandoned docklands, industrial districts and railyards.

Beginning in the 1960s, opportunistic development schemes transformed sections of the Toronto waterfront (Desfor *et al.* 1989). The efforts of municipal planners to play a formative role and to give some coherence to these interventions were largely ineffectual; the ponderous land-use planning programmes conducted by local authorities remained little more than exercises on paper. The existence of several levels of government (Fig. 10.1), together with a cluster of assorted agencies, boards and Commissions, all with some stake in one or more facets of the waterfront, created a

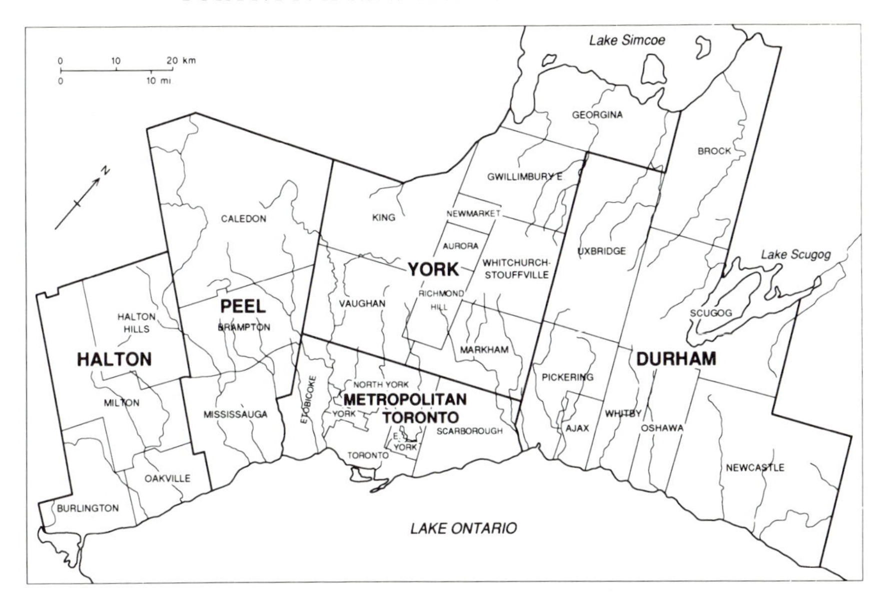

Figure 10.1 Toronto: regional governments

jurisdictional jungle that stymied most major planning initiatives.[1] At the same time, the public agencies that owned large under-utilized areas replaced or modified their early corporate goals (which included developing port, airport and railway facilities) in order to rescue their eroding financial positions by initiating waterfront redevelopment projects with private entrepreneurs (Goldrick and Merrens 1990).

In recent decades, and notwithstanding the existence of a number of major planning efforts, waterfront redevelopment in Toronto has been driven principally by market forces, weakly regulated by tangled layers of controversial, and often overlapping, governments and public agencies. The ineffectiveness of the regulatory regime, combined with a private development sector attracted by the gradual restoration of the waterfront and its profit potential, produced a pattern of redevelopment that had little in common with the desires of a public with a growing awareness of environmental degradation. A so-called 'ceramic curtain' of commercial and residential high-rises materialized on the water's edge, which, it was charged, cut off public access to the waterfront and formed a crude monolith of segregated development. During the 1980s criticism mounted, and with it the politicians' perceptions of rising political costs if the evolving trajectory continued unchecked (Stren *et al.* 1992: 142).

In the 1980s, this volatile situation provided the context for, and in an important sense precipitated, the innovative waterfront planning that is the focus of this chapter. Popular political pressure culminated in the establishment of a federal Royal Commission in 1988, and by a process of transmutation the Royal Commission became the launch pad for the adoption of a new planning paradigm for the waterfront, one that constituted an ecosystem approach. It should be emphasized, however, that the adoption and subsequent elaboration of the ecosystem planning paradigm was a gradual process. It evolved on three different fronts, sometimes simultaneously. These were: the definition and elaboration of the ecosystem approach; the development of principles to enable it to be applied; and the praxis and iteration involved in its application.

DEVELOPMENT OF THE ECOSYSTEM PARADIGM

At first, the Royal Commission investigation appeared to be a conventional 'fire-fighting' exercise set up to find remedies for several politically contentious waterfront hotspots. A basis for their solution was presented in the Royal Commission's first *Interim Report* (1989: 195–6). In the *Interim Report* – almost as an afterthought – the Commission committed itself to adopting what it termed a 'watershed approach'. This was the first indication that something more than a conventional, pragmatic remedy was contemplated. The Commission insisted that:

> . . . across the entire watershed, a 'green' strategy be devised to preserve the waterfront, river valley systems, headwaters, wetlands, and other significant features in the public interest. Such a strategy would physically link the waterfront to the river valley systems which, in turn, would be linked by the preserved headwaters areas.
>
> (Royal Commission 1989: 195–6)

When it first appeared, the significance of this particular recommendation, one among many others in the first *Interim Report*, was missed by a majority of the public affected. But three years later, in its final report, the Commission termed it the 'single most important recommendation of the *Interim Report*' (Royal Commission 1992: 6).

Following the appearance of the *Interim Report* in 1989, understanding of the embryonic watershed approach developed, emerging fully fledged as the ecosystem approach in 1990 in the Royal Commission's second interim report (*Watershed*). This represented the culmination of a substantial conceptual learning process, and it signalled the modification of traditional planning concepts by the Commission and its adoption of a new paradigm (Royal Commission 1992: 10). This new understanding was the product of many commissioned studies, consultations, hearings, workshops, and several practical experiments in the application of ecosystem principles (see below). It was informed by frameworks developed by the Healthy City Movement, the Ecosystem Charter of the Rawson Academy of Aquatic Science, the

Great Lakes Water Quality Agreement, and so on (Royal Commission 1991b: 38–43). By the time that it entered its last year of existence, the Commission could declare that: 'Fundamental to all [our] efforts was the conviction that the environment had to be the workbench on which all other aspects of the Commission's operations and conclusions would be built.' (Royal Commission 1992: 10).

The Commission defined the ecosystem at some length:

> Simply put, an ecosystem is composed of air, land, water, and living organisms, including humans, and the interaction among them. The concept has been applied to many types of interacting systems, including lakes, watersheds, cities, and the biosphere. . . . Traditionally, human activities have been managed on a piecemeal basis, treating the economy separately from social issues or the environment. But the ecosystem concept holds that these are inter-related, that decisions made in one area affect all the others. To deal effectively with the environmental problems in any ecosystem requires a holistic or 'ecosystem' approach to managing human activities.
>
> (Royal Commission 1990: 18–19)

Some of the characteristics of the ecosystem approach were enunciated by the Commission in order to tease it out and give it some substance. Essentially they elucidated the Commission's mantra, adopted from Barry Commoner, that 'Everything is connected to everything else' and that an ecosystems approach is 'based on natural geographic units – such as watersheds – rather than on political boundaries' (Royal Commission 1990: 17, 20). In *Watershed* the Commission also described a set of principles that followed on directly from the ecosystem approach to managing the waterfront. In the Commission's view, these principles should inform policies and planning by governments at all levels and provide a standard for evaluating waterfront development and management. The principles, nine in number, held that the waterfront should be: clean, green and attractive; functionally, it was to be usable, diverse, open and 'connected'; it was also to be affordable and accessible to the public. Reduced to simple, crisp and compelling assertions, the principles were elaborated at some length by the Commission in its reports, and were incorporated into later studies conducted by the Commission as criteria for analysis (Royal Commission 1991a, 1991b, 1991d).

THE APPLICATION OF THE PARADIGM

Alongside the development of its conceptual framework, the Commission had to confront the critical question of how to apply the ecosystem approach. For the Commission, no less than for all those grappling with environmental administration, no easy solution was at hand. Conventional 'rational–comprehensive' planning systems are designed to mesh with common, hierarchical administrative structures. Environmental policy-making, on the other hand, requires the integration and synthesis of many functions and specialized skills that hierarchical command systems of bureaucracy cannot deliver.

A strategy for the implementation of environmental policy required certain qualities. First, whatever was put in place had to be permanent, so that the work of the Commission could continue after the Commission itself was dissolved. Second, its jurisdiction had to be more extensive than the waterfront within the existing boundaries of the City of Toronto, of Metro Toronto, or even of the Greater Toronto Area. It had to encompass the bioregion of which the Toronto waterfront was just a part (Fig. 10.2). Third, the policy-making system had to work through the existing framework of government. By the Commission's count, there were seventy-five public agencies involved in the bioregion. But discretion being the better part of valour, it concluded that:

> No single level of government can or should be in control of [the waterfront]. The issues are too complex, cut across too many boundaries, involve too many scales and levels . . . [to be] left in one pair or even in several sets of hands.
>
> (Royal Commission 1992: 460)

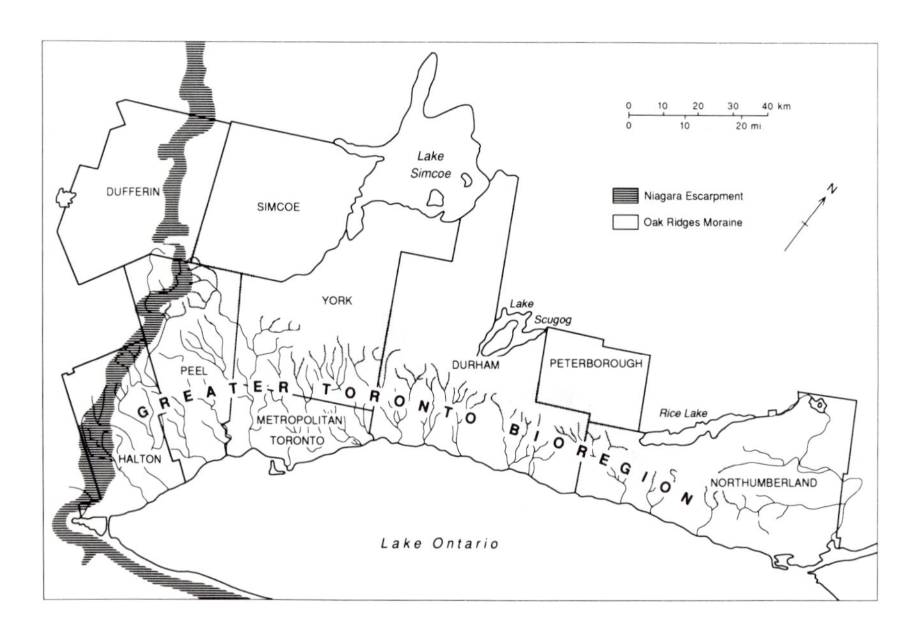

Figure 10.2 Toronto: bioregion

Therefore the challenge was to find an institutional device that could formulate complex plans with the flexibility needed to tap the diverse resources essential for their implementation. The device also required a process that incorporated features unlike those associated with linear and sequential decision-making systems typical of conventional bureaucracies. Fourth, there had to be a legislative framework compatible with the needs of both land use and environmental planning. Finally, since policy-making based on ecological principles cuts across existing institutions of representative democratic government, alternative means were required to mobilize public opinion for its support.

These requirements for the formation and implementation of environmental policy gradually crystallized and took the form of three component parts. The first was a Waterfront Regeneration Trust Agency (WRTA), which was to provide a permanent institutional vehicle and policy-making process. The second was the comprehensive reform of existing provincial legislation dealing with town planning and environmental assessment. The third component concerned the Commission's cultivation of a popular base through the encouragement of special-interest groups located throughout the bioregion. These three components comprised the framework by which the Commission hoped to implant its new paradigm for the conduct of ecologically based planning into the existing system of government. Each one of the three elements warrants some consideration.

The institutional framework

The WRTA was established by provincial statute in 1992. It followed from a lengthy process of conceptual and institutional development that started when investigations of the Toronto waterfront began in 1986, even before the Royal Commission was set up. The first few iterations of the process saw the waterfront as a series of isolated, site-specific, volatile political issues that the existing decision-making system had been unable to handle because of parochial politics, statutory competence or jurisdictional muddle. The familiar 'fire-engines' of consultants' reports and intergovernmental committees had been employed in a vain attempt to extinguish the conflagrations. With the political temperature rising, one affected party, the federal government, moved independently by deputizing one of its number, David Crombie, to negotiate settlements, and by subsequently appointing a Royal Commission, again under Crombie, to sort through the ashes.[2]

These initial problem-solving devices represented a conventional response compatible with the existing multi-tiered, fragmented structure of government. But under the impetus of public testimony and research sensitive to ecological concepts, the problematic of waterfront regeneration that confronted the federal Royal Commission was significantly altered.

Once the issue came to involve matters such as housing, transportation, health and the urban structure, the statutory competence of the federal Royal Commission was seen to be substantially inadequate. Accordingly, the Commission reached out for access to additional powers vested in the provincial government and municipalities, attempting initially to mobilize them in a voluntary Intergovernmental Working Committee, and then negotiating with the federal and provincial governments to establish a joint Royal Commission. This union brought to bear the combined powers of the two governments, thereby permitting an unconstrained examination of the ecological, social and economic requirements of environmental regeneration. A second consequence of this union was that waterfront planning was linked to development in the bioregion area, of which the waterfront was merely the edge. Explicitly coupling waterfront planning to the pace, intensity and nature of the upstream built environment and the impact of its development on the biosystem placed the waterfront in a much larger, more intricate context.

The series of transformations did not end with this new joint body. Once the problems that had been the particular concern of the federal government appeared to be partly resolved, the government dropped out of the partnership in 1992 and the Commission became an exclusively provincial commission of inquiry, but one which maintained informal links with the federal government. This series of transformations provided the policy-making process, represented by the Commission, with access to the powers required to formulate comprehensive plans that incorporated principles of environmental maintenance and regeneration.

The new provincial Commission's form of policy development was compatible with the scope and areal extent of the problems relating to waterfront development as it had come to be defined. The provincial Commission was not bound by existing governmental jurisdictions, neither was it confined by existing functional conventions of bureaucratic organization. Furthermore, it had a statutory mandate to work within the broad range of provincial powers, and its situation as an independent, free-standing policy centre gave it the capability to formulate recommendations that were regional in scope and multi-functional in character. Its independent status also gave it leave to operate within the bioregion, linking combinations of public bodies as the adoption and implementation of its recommendations required. Thus the penultimate manifestation of the transformation process was, in theory at least, one that approximated more closely than its predecessors to a policy-making vehicle that had the scope and powers required by the new planning paradigm.

The provincial Commission, however, was only a temporary device. The expiration of its mandate would simply mean that its concepts and recommendations would be left to existing planning practice and to the not-so-tender mercies of the existing machinery of government. By the time that the Royal Commission had issued its final report (*Regeneration*, 1992), the last phase of the transformation had occurred. A Waterfront Regeneration Trust Agency was established by statute as a non-profit-making agency.

The Agency is empowered to exercise its functions throughout the entire Toronto bioregion, an area that extended from Halton, west of Toronto, to Northumberland in the east, a total of 250 km, and northward in a 50-km arc to the Oak Ridges Moraine. This area embraced all or parts of many local and regional governments, as well as numerous special-purpose bodies. The powers conferred by the province on the Agency are very general but potentially extensive. The Agency is intended to take the lead in public consultation and to advise the Minister of the Environment on the use, disposition, conservation, protection and regeneration of lands adjacent to Lake Ontario (to an unspecified depth); to involve itself in environmental, land-use and transportation planning; to expedite decisions on waterfront issues; to work with all levels of government, community groups and the private sector, in order to achieve job creation, economic development and a healthy waterfront; and to co-ordinate programmes and policies of the provincial government and its agencies relating to waterfront lands. The extent and definition of these powers probably depends on the manner in which they are exercised. Because the Agency is not given direct powers to implement its preferred programme, these are to be negotiated with units of existing governments having operational responsibilities. Therefore, the range of the authority wielded by the Agency will emerge, as the Minister suggested, from 'a blend of leadership, consultation, co-operation, initiative and diplomacy' (Ontario Ministry of the Environment 1992: 2). Furthermore, its powers of persuasion are generously reinforced by provisions that permit it, with the approval of the Ontario Cabinet, to acquire property, borrow money and exercise any powers 'that are necessary or expedient for carrying out its functions' except as limited by the Act (Ontario Ministry of the Environment 1992b: 2). Few limitations are stated.

Through a circuitous process, what originated as a conventional problem-solving exercise focused on several site-specific 'hot-spots' on the Toronto waterfront was transformed into a instrument incorporating many of the features required for ecologically based planning and policy-making. However, whereas the WRTA provided an institutional vehicle for analysis and policy-making, in practical terms the manner in which it could conduct its work had to be determined. Its status as a special-purpose body affords a relative autonomy from existing institutions of government, but it has no explicit executive powers. Deficiencies of conventional planning practice aside, the Royal Commission identified several major obstacles to implementing an ecosystem approach. One was the rigidity of bureaucratic systems, another was the fragmentation of governmental jurisdictions and a third was the functional isolation of sectoral interests within and between public and private organizations:

> They combine to create a high degree of paralysis that pervades our systems of governance, and makes it difficult, if not impossible, to make sound, integrated decisions.

It was the view of the Commission that '*if we want to improve the kind of decisions we make*, we are going to have to change the *way* we make decisions' (Royal Commission 1992: 46).

The practical challenge confronting the Commission, therefore, was to find an approach that would permit it to work through existing government systems, yet at the same time deal with the multi-disciplinary, cross-sectoral, and multi-jurisdictional nature of problems rooted in the ecological, social, economic and political systems with which it had to contend. The response of the Commission was twofold. On the one hand, it adopted the role of a catalyst or agent of change, attempting to combine resources, aggregate interests and design strategies, and to co-ordinate responses and stimulate action. The technique adopted by the Commission was known as 'round-tabling', which features a concurrent rather than a consecutive planning process. Great store was put by this consensus-building device. The Commission described it as 'the real key to the public administration of the waterfront' (Royal Commission 1992: 461). It was tested and used extensively by the Commission in its work.

A variation on the technique is also being employed to make operational the WRTA's 'agent of change' role. The province adopted the Commission's recommendation that the task of fashioning strategies for eco-based planning should be transacted by the WRTA through devices known as Waterfront Partnership Agreements. The 'stakeholders' involved in particular issues are brought together by the Agency to develop, by consensus, agreements that become binding commitments on organizations, public and private alike. This technique proposes the drawing together of relevant resources and decision-makers from affected organizations and constituencies identified not by existing government jurisdictions but according to the incidence of environmental impact.

This approach is related to environmental dispute resolution techniques which have been practised in the United States and Canada for the past decade. These are described not so much as methods of analysis but as political procedures, their claim to legitimacy resting on the pluralist assumption that if the process is open and consensus-based, then the resulting policy is, by definition, a good one (Amy 1990: 70–5). Whatever its merits, the Commission and the provincial government embraced this approach as a means which, it was claimed, would cut through the bureaucratic stasis and jurisdictional gridlock that were identified as major constraints on the operation of ecosystem-based policy-making.

With a new, permanent institutional vehicle in the form of the WRTA in place, and with appropriate operational processes to animate it, the next task for the Commission was the development of a legislative framework to secure the integration of environmental and conventional land-use planning.

Statutory reform

As the Royal Commission moved towards the adoption of its ecosystem approach, deficiencies in the regulatory environment for land-use and environmental management gradually became apparent. The Province of Ontario has two sets of administrative and quasi-judicial procedures broadly related to regulating land use, with essentially no provision for co-ordination between them. One, the Planning Act, is local in jurisdiction, reactive and regulatory in application, and functions outside any comprehensive framework of provincial land-use policy. Moreover, it contains only vague references to 'the protection of the natural environment' (Ontario 1989, Section 2). The second statute, the Environmental Assessment Act, while unique in the breadth of its definition of 'environment', has been applied only rarely to private-sector development, and has been applied in varying fashion in the public sector (Royal Commission 1991c: 29). The two statutes were developed in isolation from one another and do not work together to provide an integrated system for environmentally sound planning.

The province eventually complied with the Commission's recommendation to set up a commission of inquiry to reform fundamentally the land-use planning system in the province (Royal Commission 1991c: 77). The inquiry, named the Commission on Planning and Development Reform in Ontario, presented a draft version of its final report late in 1992. While the terms of reference assigned to the inquiry did not once mention the word 'environment', the Royal Commission's final report, *Regeneration*, claimed that while the scope of the inquiry was not as broad as recommended, it would consider incorporating environmental concerns into the Planning Act. The draft final report succeeds in achieving this goal (Ontario Commission on Planning and Development Reform 1992).

Implementation of recommendations made by the Planning and Development Reform inquiry will require legislative changes by the provincial government. But in so far as it was able, the Royal Commission proposed, advocated and initiated a process that could lead to the incorporation of land-use and environmental matters into a statute to facilitate integrated planning and policy-making.

Popular base

The third and final component of the framework through which the Commission hoped to implant its new paradigm in the existing system of government comprised an institutionalized popular base. Throughout its existence, the Commission consulted very widely through visits, hearings, consultations and the like. This reflected the style of the Commissioner, David Crombie, and the Commission readily acknowledged its debt to the many deputants whose emphasis on environmental concerns had played an important role in sensitizing the Commission to the issue:

> . . . dozens of deputants delivered the same message: by all means sort out the issues of Harbourfront and the Harbour Commissioners, but help us find out how to make our lake publicly accessible, fishable, drinkable, and swimmable. This cannot happen while the rivers that empty into the lake are contaminated, the air that connects to it is dirty, the groundwaters polluted, and the soils through which they pass contaminated.
>
> (Royal Commission 1992: 2–3)

Clearly, the Commission was being told that environmental issues should be its paramount concern. Implicit in these responses was the plea that planning and development of the waterfront should be based on what was, at that time, the fairly novel and undeveloped concept of the ecosystem. But the Commission recognized that between the sentiments of citizens and its own objectives and policies lay the mediating presence of entrenched interests and bureaucratic inertia. The principals of the Commission expressed their resolve that its work should not end up gathering dust on the shelf of a provincial ministry. They felt that the surest way to prevent this was to develop a constituency that had a clear interest and a personal stake in the ecosystem approach (Doering 1991: 20, 62). Accordingly, the Commission once again recommended that the province adopt measures to achieve this, while, at the same time, it initiated the constituency-building process. Its recommendation was that a citizens' coalition should be formed, with the assistance of the province, to provide research and advocacy and to address issues that crossed traditional jurisdictional boundaries. Moreover, the coalition should receive funding supplied to intervenors, and other assistance (Royal Commission 1990: 85). For its own part, the Commission helped to form a group known as Citizens for a Lakeshore Greenway, and generously promoted the work of other environmental groups working on specific projects.

Having concluded that it would adopt a new paradigm, the ecosystem approach to planning and urban policy-making, the Commission was faced with the need to construct a framework composed of institutions, regulations and a popular base, if its concepts and practices were to be placed permanently within existing institutions of government. The WRTA, the Commission on Planning and Development Reform, and the encouragement of citizen advocacy were three principal components of its approach. The three components, like the paradigm itself, were the product of imaginative thinking, but they were also the result of a process of praxis in which the Commission engaged.

PRAXIS

There were several reasons for the Commission's emphasis upon praxis. The ecosystem approach, particularly with respect to its application in cities, is new and requires experimentation. Also the Commission knew there was pressure to deal with some politically sensitive waterfront problems, and that its original purpose had been to resolve these. Furthermore, the participative nature of the Commission's work constantly brought it face-to-face with activists and public officials who addressed the Commission from a very practical and compelling perspective. Finally, the Commission's Chair, Crombie, was temperamentally inclined towards activism, problem-solving and intervention. Therefore, while the Commission grappled with the conceptual problem of the ecosystem approach, its work was firmly rooted in practical, current issues. In the words of the Commission, 'the work ranged from theory to practice, policy to program' (Royal Commission 1992: 10). As a consequence of this, and probably to its great advantage, the Commission was characterized by features of praxis, advocacy and facilitation.

Three projects tackled by the Commission between 1989 and 1992 illustrate its experience. They were said to 'cover the most important issues on the waterfront: environment, transportation, and land use' (Royal Commission 1992: 304). The first was a major environmental audit undertaken in late 1989 on 567 ha of strategically located, heavily polluted land, known as East Bayfront/Port Industrial Area, lying adjacent to the core of the City of Toronto. The work was recommended by the Commission in its first *Interim Report* (1989) and accepted by the governments of Canada and Ontario. The audit was a pioneering attempt to apply the ecosystem approach and to test the efficacy of the Commission's preferred 'consensus-building' model of decision-making (Royal Commission 1992: 391–2)). In the first phase of the audit, a large number of 'stakeholders' was assembled to gather information on aspects of the ecosystem. Technical papers were produced, covering aspects of the terrestrial, aquatic and atmospheric environments of the site. Extensive public hearings were held to provide opportunities for discussion and feedback. The second phase tackled the less familiar and more subtle relationships among the various elements of the environment: the atmosphere, built heritage, ecosystem health, hazardous materials, natural heritage, soils and groundwater, and water and sediments (Royal Commission 1991b: 117).

Though the audit was recognized to be preliminary in many respects, it did provide, in substantive terms, a better understanding of the structure and function of the ecosystem, and, in terms of technique, it provided a template for the application of such analysis. The government of Canada proclaimed that this audit was a multi-governmental partnership, which it regarded 'as a model of the way in which it can work with other levels of government to achieve common objectives through the application of an ecosystem approach' (Canada 1990: 32).

The second and third projects undertaken by the Commission also experimented with the application of ecosystem principles but incorporated strong policy components as well. Both involved sites located in the Central Waterfront of the bioregion, in an area regarded by the Canada–US International Joint Commission as having clean-up problems as complex and difficult as any in the Great Lakes (Royal Commission 1992: 303).

The *Toronto Central Waterfront Transportation Corridor Study* (Royal Commission 1991d) contemplated the barrier of an elevated expressway and the multiplicity of railway lines, which virtually cut off the core of Toronto from the shore of Lake Ontario. In the eyes of the Commission, the modification of this barrier would determine whether the people of Greater Toronto would have a waterfront integrated with its central area, or whether the two would remain essentially separate (Royal Commission 1992: 306–7). The study was based in part on the Commission's earlier work, particularly the audit of East Bayfront. The aim was to apply the ecosystem approach, utilizing the Commission's nine principles, so as to strike a balance between the dual functions of the study area as a 'place' and as a 'corridor'. The term 'corridor' referred to its transportation role, and 'place' connoted its residential and recreational qualities as a site that was clean, green, usable, diverse, connected and attractive (Royal Commission 1992: 318–19). The study sought to achieve such a balance, between the strategic regional function of the area and the prevailing urban-design goals of diversification and intensification of land uses in the Greater Toronto Area.

Another project dealt with a complex, central site of 500 ha located on Lake Ontario. Known as Garrison Common, the area features mixed uses dominated by declining public recreation and exhibition facilities, and is bisected by transportation routes (see Chapter nine, Fig. 9.8). The Commission again recruited a study team, which, applying the ecosystem approach, examined the site in terms of its relationship to the biophysical and human environments and the natural, social and economic consequences of its development.

One of the principal challenges of the site concerned the highly fragmented patterns of land ownership and the means by which an integrated plan could be realized. Four levels of government and several special-purpose bodies were involved, together with private owners. Individual owners were preparing development plans in isolation from one another within the framework of the conventional planning regime. The task of the Commission was to encourage redevelopment, but to do so in the context of the ecosystem approach. The Common was to be incorporated into a 'green corridor' along the lakefront, with provision made for the enhancement of historic sites and natural features, the development of new recreation and tourist facilities, and the integration of existing residential neighbourhoods with the site and the adjacent waterfront. Given the complexity of ownership, the remedy proposed by the Commission was that a Waterfront Partnership Agreement be struck among the interests that were involved, under the aegis of the WRTA (Royal Commission 1991a: 91).

Like the East Bayfront/Port Industrial Area, and the Transportation Corridor, Garrison Common is an extremely difficult site in which to apply the new ecosystem paradigm. All of the areas lie in an uncompromisingly urbanized zone in which the natural environment has been relentlessly supplanted by successive waves of development. In addition, each scheme deals with the key issues of environment, transportation and land use. Consequently, they not only served as a challenging 'test-bed' for the Commission's approach but also provided it with ample opportunity to explore and refine its ecosystem approach.

The Commission also engaged in two other forms of practical work. These enriched its understanding of ecosystems, enhanced its legitimacy in the eyes of its governmental masters, and generated support and understanding of its work among the general public. The first of these practical works concerned the troublesome areas that had engaged the Commission's attention from the outset. One of these was Harbourfront, the 1992 election gift of the federal government, which had started out as a large urban park but by the mid-1980s had become a contentious 'ceramic curtain' of high-density, luxury buildings (see Chapter nine, Fig. 9.7). Another was the East Bayfront/Port Industrial Area, which comprised derelict industrial sites. Both required immediate attention. Harbourfront was the more intractable area – because its built-up condition offered little room to manoeuvre. After many public hearings and considerable effort, the Commission was able to negotiate agreements with three levels of government, which led ultimately to the reorganization of Harbourfront and, in essence, 'greened' portions of its site in a manner that appeared to satisfy its critics. To achieve this, the Commission acted as an aggressive advocate and facilitator seeking a pragmatic settlement designed more to achieve a 'satisficing' solution than one employing an elegant, ecosystem-based approach.

The East Bayfront and Port precincts were under-utilized and clearly in transition. Here, the Commission could apply its favoured planning approach through the introduction of a ground-breaking environmental audit (see p. 232). The audit did not produce a specific plan. But in addition to specifying stresses in the ecosystem and analysing their interaction, it provided an extensive set of criteria for evaluating future development and an integrating set of goals and implementation strategies. These were adopted by key levels of government and their pursuit was handed to the Waterfront Regeneration Trust Agency for development.

The second activity that engaged the Commission took it beyond the central waterfront of Toronto to suburban communities and to municipalities along the 250-km shore of Lake Ontario at the southern edge of the bioregion. Public hearings were held in many of these areas, and the Commission promoted discussion of waterfront regeneration and intervened directly in issues that affected the integrity of the waterfront and upstream watersheds. Despite some wariness, bordering on hostility, to uninvited advice, by people from Toronto (reported in the local media), the Commission was surprisingly successful in implanting its ecosystem approach in the minds of citizen groups and the strategies adopted by planning officials (*Oshawa Times,* 22 October 1989; Royal Commission 1992: 269, 462). In addition, with the supportive provincial government, the Commission was able to facilitate considerable intergovernmental co-operation on planning issues according to their bioregional incidence, rather than traditional institutional boundaries.

The Commission conducted its work in a fairly unusual fashion. It was a thinker, experimenter, analyst and executor. Its proclivity to deal with practical issues while developing new conceptual approaches was the product of temperament, practical political realities and conviction. The work of the Commission on the audit of the East Bayfront/Port Industrial Area, the Transportation Corridor and Garrison Common, along with its activities outside the Metropolitan Toronto area, introduced the new paradigm to governments and the public, while it initiated communication between previously isolated planning interests. The rhetoric must be measured against the reality, but governments ranging from the Province of Ontario to Metropolitan Toronto, other regional governments, and second-tier municipalities have adopted the ecosystem approach in recent iterations of their official plans and subsidiary documents. Whether this represents fashionable rhetoric or firm commitment remains to be seen.

CONCLUSION

We conclude with a number of observations suggested by the analysis. The first concerns the lack of congruence between conventional, hierarchically structured organizations and those that are suggested by the nature of ecosystem-based urban planning.

Public organizations can be regarded as gate-keepers which reflect the values and particular interests of those given preferred access to decision-making, or those excluded or discouraged from seeking entry. There is no better example of this proposition.

Contemporary structures of government operate from a growth-driven paradigm, largely blind to environmentally defined constraints (Rees and

Wackernagel 1992: 6). For their part, municipal governments have been characterized as 'growth machines' and have been designed accordingly (Logan and Molotch 1987). They are functionally structured bureaucracies. Health, housing, public works, social services and transport, for example, provide services in support of development that originates in the private sector. Even one of the principal municipal 'service' functions – planning – is concerned primarily with processing development applications. In any plan-making in which they may engage, city planning organizations typically have little influence on the work of entrenched 'line' departments. The entire system lacks flexibility, is bound within arbitrarily defined political jurisdictions, operates linearly and often at a snail's pace, seeks to define issues in ways that fit existing organizations (thereby validating their existence), and is incapable of synthesizing complex variables that fail to conform to its rigid structure.

Clearly, the planning/policy-making paradigm of the ecosystem approach is incompatible with the traditional bureaucratic pattern. Its values and its interests simply cannot gain organizational purchase and achieve legitimacy. New ideas and new ways of doing things need room within existing organizations if they are to become etched in the policy-making process. Flexible, problem-solving organizational forms seem to be needed, rather than hierarchies constructed around routine tasks. The Toronto waterfront planning exercise evolved from the latter to something that looks a little like the former, although in a partial or hybrid form. The eventual broad jurisdictions of the Royal Commission and WRTA, defined by environmental criteria, give greater flexibility to the identification of affected interests and access to the resources needed to tackle specific issues according to the requirements of the ecosystem approach. Whether the WRTA, as a pragmatic institutional response to a static system, has the authority to go beyond that and secure the implementation of policy by existing governments is a question that cannot yet be answered.

The second observation speaks to the first by illustrating the extreme difficulty of achieving institutional innovation in a complex federal system such as the Canadian government. Writing two decades ago, Harvey Lithwick observed that 'The growing gap between the technology of problem creation and the technology of problem solution is perhaps the best explanation for the failure of public policy for the cities' (Lithwick 1971: 17). In a sense, the ontogeny of the Royal Commission and the WRTA is about the search for a means to reduce that gap; to discover a device capable of concentrating, within one policy centre, powers that are diffused throughout four levels of government and many special-purpose bodies and popular groups. None was capable individually of fashioning such a policy framework, or of extracting from others, and reconstituting in a novel form, the powers needed to integrate environmental and planning concerns. Attempts at independent action by the City of Toronto, the Metro government and the government of Canada produced few results. Co-operative measures, working with the Intergovernmental Working Committee, similarly were ineffective. The eventual formation of a joint Royal Commission, however, created a force that was not bound by established practice and institutional constraints. The Commission had the independence and the cachet needed to formulate a new, comprehensive planning paradigm and to propose how the powers dispersed throughout the governmental system should be deployed to achieve its aims.

Therefore, through a series of iterations, a policy centre evolved gradually and, in its final form, was institutionalized as the WRTA under the aegis of the provincial government. Conceivably, the innovation in the 'technology of problem solving' represented by the Agency could have been introduced unilaterally by the province. However, the inertia inherent in the complex system of public authorities in the Toronto bioregion was unlikely to yield to anything but the momentum created by a strong, external influence like the Royal Commission.

The final observation addresses the socio-political context of environmental planning. Popular pressure had much to do with the initiation of the process that produced what might be described as a prototype of a new institutional structure that could provide inspiration, impetus and co-ordination to the urban-planning project (Stren *et al* 1992: 142). But what does the new structure contribute to the goal of environmental sustainability? Exploration of this issue is not the subject of this study, and an evaluation would be premature. However, it is clear that since cities are the nexus of consumption and waste creation, a balance must be struck between the cities themselves and their ecological carrying capacity and that of their appropriated hinterlands. As this goal is inimical to current concepts of production and distribution, which are predicated on constant growth and differential consumption, the path to sustainability lies more through behavioural and ideological modification than it does through the remedy of institutional innovation as implied by the Commission (Royal Commission 1992: xxi). Appropriately structured administrative systems can convert popular opinion into public policy. But institutional innovation will not, in itself, lead to the regeneration of cities in a manner that is compatible with environmental sustainability. It is a necessary step, but not a sufficient one.

In embracing an ecosystem-based planning paradigm, the Royal Commission promised much. It based its approach on an understanding of natural processes at work in bioregions (Royal Commission 1992: 41), an understanding that was explicitly derived from Kirkpatrick Sale's 1985 treatise, *Dwellers in the Land: The Bioregional Vision*. However, the importance of natural processes is only a part of the context of urban planning. In a recent appraisal of environmentalism, Sale himself emphasized the inherent limitations of efforts that fail to recognize that environmental problems are 'inevitable byproducts of an economic system based on the imperative of growth . . . and governments designed to protect it'. Efforts to effect change that ignore these considerations 'tend to confine themselves to piecemeal reforms rather than structural changes and to isolate such problems and their solutions from what might be called a political context' (Sale 1993: 94). Its bioregional rhetoric notwithstanding, the Royal Commission's vision was a blinkered one.

NOTES

1. Under the terms of the division of powers in the Canadian Federal State, provincial governments are assigned direct responsibility for most matters concerning the built environment, including provisions for property and human rights. Aspects of these matters are delegated to municipalities by the governments of each province (Sancton 1992). The municipal structure of government in the area under consideration here comprises five regional governments, each one of which is organized as a federation. Lower-tier municipalities include cities, towns, villages and rural authorities. Within all of these there exist a variety of special-purpose bodies that enjoy varying degrees of autonomy. Powers conferred by the provincial government are divided between units in each federated regional government. Those of a regional nature reside with the upper tier, while local ones are vested in municipalities in the lower tier. Some powers are shared, including those relating to the regulation and use of land, with the province standing as the final arbiter. In the Metropolitan Toronto area, the upper-tier regional government, referred to herein as the Metro government or Metro, shared power with the City of Toronto government, or the City, and five other lower-tier units (Fig. 10.1).
2. David Crombie had been Mayor of the City of Toronto from 1973 to 1979, and subsequently served in the Cabinet of the federal government for a number of years. His political effectiveness and great popularity were built on a set of conservative beliefs that emphasized values such as tradition, preservation, stability, stewardship and community. His approach as Commissioner of the Royal Commission and then as Commissioner of the WRTA reflected these values.

BIBLIOGRAPHY

Amy, D. J. (1990) 'Decision techniques for environmental policy: a critique', in R. Paehlke and D. Torgerson (eds), *Managing Leviathan: Environmental Politics and the Administrative State*, Peterborough: Broadview Press.

Canada (1990) *Canada's Green Plan for a Healthy Environment*, Ottawa: Minister of Supply and Services.

City of Toronto, Planning Board (1962) *The Core of the Central Waterfront*, Toronto: City of Toronto.

Commoner, B. (1972) *The Closing Circle*, New York: Bantam Books.

Desfor, G., Goldrick, M. and Merrens, R. (1989) 'A political economy of the waterfrontier: planning and development in Toronto', *Geoforum* 20(4): 487–501.

Doering, R. (1991) untitled article, *Manager's Magazine* (December): 17–20, 62.

Goldrick, M. and Merrens, H. R. (1990) 'Waterfront changes and institutional stasis: the role of the Toronto Harbour Commission, 1911–89', in B. S. Hoyle (ed.) *Port Cities in Context: The Impact of Waterfront Regeneration*, Southampton: 1990 Transport Geography Study Group Institute of British Geographers.

Hahn, E. (1991) *Ecological Urban Restructuring: Theoretical Foundation and Concept for Action*, Berlin: Research Professorship Environmental Policy Science Centre.

Lithwick, N. H. (1971) *Urban Canada: Problems and Prospects*, Ottawa: Minister of State for Urban Affairs.

Logan, J. R. and Molotch, H. L. (1987) *Urban Fortunes*, Berkeley: University of California Press.

Merrens, R. (1988) 'Port authorities as urban land developers: the case of the Toronto Harbour Commissioners and their Outer Harbour project, 1912–68', *Urban Historical Review* 17/2: 92–105.

Ontario (1989) 'Planning Act', Toronto: Ontario Ministry of the Attorney General (unpublished paper).

Ontario Commission on Planning and Development Reform (1992) *Draft Report*, Toronto: Queen's Printer for Ontario.

Ontario Legislative Assembly (1992) 'Bill No. 1: An Act to Establish the Waterfront Regeneration Trust Agency', Toronto: Queen's Printer for Ontario, 25 June 1992 (unpublished paper).

Ontario Ministry of the Environment (1992) Press release on Act to Establish Waterfront Regeneration Trust introduced by Environment Minister Ruth Grier, Toronto, 6 April 1992.

Oshawa Times (1989) '"Greater Toronto Area" a fiction', *Oshawa Times*, 22 October 1989: 6.

Rees, W. E. and Wackernagel, M. (1992) 'Appropriated carrying capacity: measuring the natural capital requirements of the human economy', paper given at the

Second Meeting of the International Society for Ecological Economics 'Investing in Natural Capital', Stockholm, 3–6 August (unpublished paper).
Royal Commission on the Future of the Toronto Waterfront (1989) *Interim Report*, Ottawa: Minister of Supply and Services Canada.
—— (1990) *Watershed*, Ottawa: Minister of Supply and Services Canada.
—— (1991a) *Garrison Commission: Preliminary Master Plan*, Toronto: Berridge, Lewinberg, Greenberg, Ltd.
—— (1991b) *Pathways: Towards an Ecosystem Approach*, Ottawa: Minister of Supply and Services Canada.
—— (1991c) *Planning for Sustainability: Towards Integrating Environmental Protection into Land-Use Planning*, Ottawa: Minister of Supply and Services Canada.
—— (1991d) *Toronto Central Waterfront Central Transportation Corridor Study*, Toronto: IBI Group.
—— (1992) *Regeneration. Toronto's Waterfront and the Sustainable City: Final Report*, Ottawa: Minister of Supply and Services Canada.
Sale, K. (1985) *Dwellers in the Land: The Bioregional Vision*, San Francisco: Sierra Club Books.
—— (1993) 'The U.S. green movement today', *The Nation*, 19 July: 92–6.
Sancton, A. (1992) 'The municipal role in the governance of Canadian cities', in T. Bunting and P. Filion (eds) *Canadian Cities in Transition*, Toronto: Oxford University Press.
Stren, R., White, R. and Whitney, J. (eds) (1992) *Sustainable Cities*, Boulder: Westview Press.

11

AMSTERDAM: THE WATERFRONT IN THE 1990s

Arne Bongenaar and Patrick Malone

INTRODUCTION

Amsterdam offers a remarkable opportunity for waterfront development. There is roughly 630 ha of development land along the River IJ immediately to the north and to the west of the city centre, of which 326 ha are under water. Discussions about redevelopment began in the early 1980s, but it took five years to produce a plan for the area (which was still subject to dispute). In 1985, a strategic decision was made to parcel the land available for redevelopment into two separate projects. Following an earlier strategy, the eastern port area (Oostelijk Havengebied) was eventually designated for the development of roughly 8,000 dwellings (Fig. 11.1). This part of the project was set to provide rented and private housing (ranging from social to expensive dwellings) in a new residential district (Municipality of Amsterdam 1985a, 1985b, 1989 and 1991b). The second area, the more centrally located IJ-oevers (literally: the border of the River IJ), was designated for office development, retail space, housing and other public and cultural facilities (Fig. 11.2).[1]

The division of the area flanking the IJ between two separate projects, each with its own development programme, has proved important to the success of the housing area. As a separate entity, this has not suffered the fate of the IJ-oevers or central area where plans for office and commercial development have, as yet, failed to materialize.

This chapter explores the background to the IJ project and the reasons for the failure of initial plans for office development in the central (IJ-oevers) area. Second, it examines the context for office development and the position of the IJ-oevers project in relation to the Dutch office market and the economy of Amsterdam. It also covers the future prospects for office development on the IJ and recent efforts to adapt plans to fit local constraints and market opportunities.

The failure to develop the IJ-oevers area means that it may now absorb a revised programme that will take the redevelopment of Amsterdam's waterfront into the next century. However, the potential for development hangs

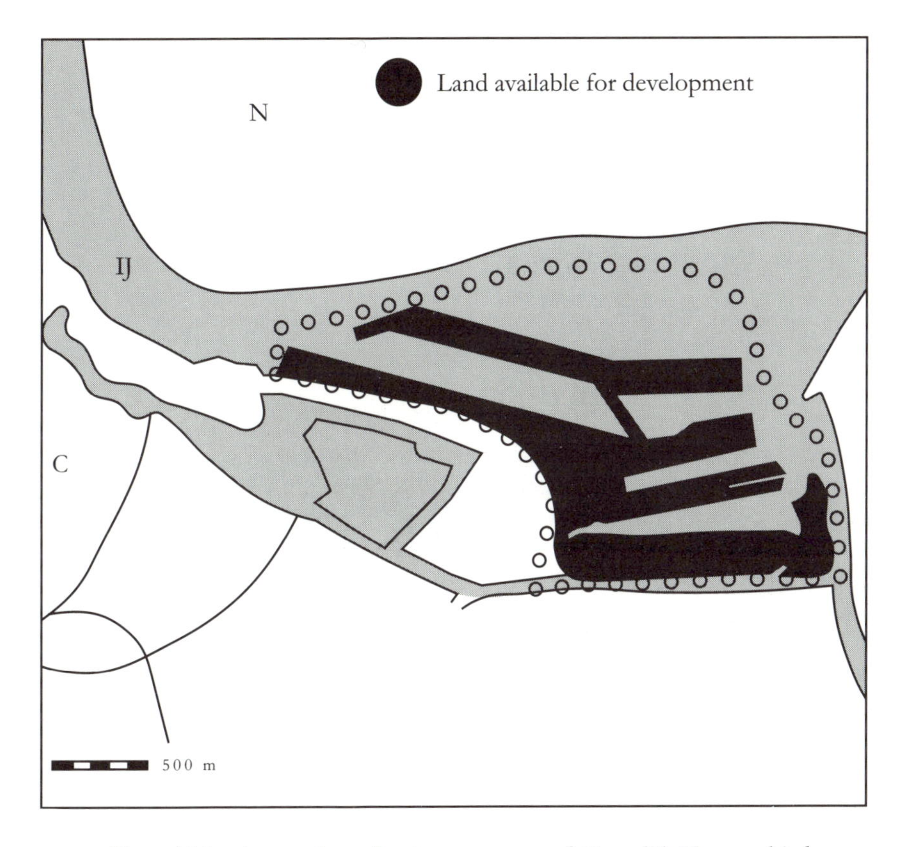

Figure 11.1 Amsterdam: development area of Oostelijk Havengebied
Key: N = North; C = Central

on the commitment of public and private interests and the capacity of those interests to attune the project to real sources of demand and the specific conditions for development in the IJ-oevers area in the 1990s. In this respect, the initial failures of the IJ-oevers project may hold lessons for developments that are economically centred on office space. It is useful to examine the reorientation of the IJ-oevers project in terms of developments where the office component (initiated in the 1980s) has been curtailed or abandoned in the face of recession in local and international markets.

BACKGROUND TO THE DEVELOPMENT

Development land along the River IJ has become available through the typical pattern of decline and relocation of port facilities (see for example

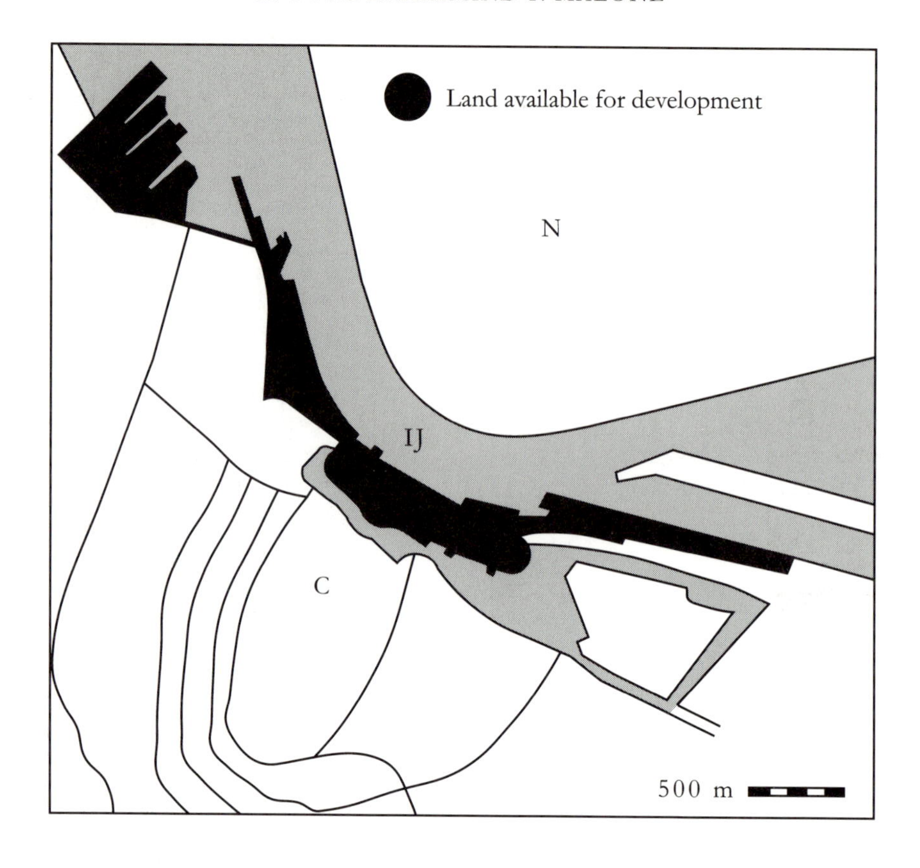

Figure 11.2 Amsterdam: development area of Zuidelijke IJ-oevers
Key: N = North; C = Central

Hoyle *et al.* 1988). The process of decline began in 1876 when the North Sea Canal was completed to the west of the old port area of Amsterdam. This provided a considerably shorter link to the sea, but it signalled the marginalization of the old port that had played such a prominent role in the Dutch 'Golden Age'. Until 1927, new quays and manufacturing plants were still constructed along the River IJ to the east of the inner city (Municipality of Amsterdam 1989). The Central Railway Station was built on an artificial island in front of the old port towards the end of the nineteenth century. In 1912, however, the municipality adopted a policy that directed further expansion of the port westward along the North Sea Canal. This accelerated the development of new port facilities to the west until, in 1978, the municipality actively relocated companies from the east to the western port area. The eastern area was earmarked for redevelopment under a 'compact city' policy that was later incorporated into national and regional planning and is still supported by the Dutch government (Ministerie van Volkhuisvesting, Ruimtelijke Ordening en Milieubeheer 1990a).

The shift westwards from the old port was reflected in a general suburbanization of Amsterdam's economy. Between 1960 and 1995 the number of jobs in Amsterdam's inner city decreased from 160,000 to less than 80,000, whereas the total number of jobs in the Amsterdam region fell from 360,000 to 300,000 (de Hen *et al.* 1986; Municipality of Amsterdam 1980–90). Building height restrictions, conservation policies, poor road access and relatively low levels of parking have all contributed to the relocation of office activities from the centre to the outskirts. Office locations near Schiphol Airport and along the subway and circular motorway became particularly popular (Table 11.1).[2]

The extent of peripheral office employment now concerns the city's authorities. In 1983, average spending in retail outlets and restaurants in the city centre was calculated at roughly Dfl10–12 (or roughly £5 per employee per day) (Hart 1984). Although these figures have not been updated, it is generally accepted that suburbanization has gradually reduced spending by employees in the city centre. The characteristic inner-city land-use mix of business activities, housing, shopping and tourism is also threatened. This has led to the prospect of an inner city dominated by tourism and housing – a prospect that is presented as the 'Venice scenario'.

In the early 1980s, recognition of new opportunities in the redundant wastelands along the River IJ emerged alongside a growing awareness of the declining economic strength of the inner city. In 1982, the municipality

Table 11.1 Amsterdam: office locations and average prime rents

Location	*Prime office ('000,000 m^2)*	*Prime rents (Dfl per year)*
Centre	1.4	300
Teleport	0.6	275
South	0.8	425
South-east	0.8	275
Rest	1.3	250
Total Amsterdam municipality	4.9	—
Schiphol Airport	0.2	600
Hoofddorp	0.3	250
Rest	0.9	250
Total Amsterdam region	6.3	—

Note: Data refer to prime office space. Rents are averaged over one/two years 1993/1994. Exchange rate (March 1996): 2.44 Dfl = £1; 1.57 Dfl = US$1.

held an urban design competition for the Easterdock. This was generally regarded as the area that could spearhead the process of redevelopment on the waterfront. At that time it was already clear that Oostelijk Havengebied, or the eastern port area, was better suited to housing.

THE EASTERN PORT AREA

The eastern port area (Oostelijk Havengebied) covers a massive 300 ha, of which 132 ha is land. In 1978, when the decision was taken to redevelop the area, more than half of it was already owned by the municipality.

The decision (taken in the mid-1980s) to designate the area for housing was made mainly in response to housing needs, but it also reflected the fact that the area was not suitable for office development or inner-city functions. Designation for housing raised a number of questions, and particularly the issue of costs (Bongenaar and Veerkamp 1987). As a former port area, the site required extensive restructuring, the purchase and removal of Dutch Railways track, the treatment of heavily polluted soil, and either demolition of existing buildings or their integration into new plans. Moreover, a number of leaseholds (some due to expire in the year 2020) needed to be bought or renegotiated, and substantial removal fees paid to relocating companies.

In line with other acclaimed redevelopment projects in Amsterdam, 'integration' was a key factor in development proposals for the eastern port area. Attention was given to the integration of land uses and the links between the area and the rest of the city. The project was heavily subsidized by central government under a scheme for the revitalization of inner cities (Subsidie voor Grote Bouwlocaties). However, central government also reduced the number of subsidized dwellings in the project to 43 per cent of the total of roughly 8,000 units (Municipality of Amsterdam 1989). Initially, the municipality wanted a larger proportion of social housing, but later welcomed the reduction imposed by central government. After a century of socially orientated housing policies, Amsterdam's housing stock is relatively lopsided. Approximately 90 per cent is rented housing, and 75 per cent is set at a low rent (Municipality of Amsterdam 1980–90). Thus, the municipality must consider the long-term effects of the relocation of Amsterdam's wealthier citizens to the suburbs.

The housing area will cover 123 ha, and the overall density for housing in the eastern port area was set at sixty-three dwellings per hectare (Municipality of Amsterdam 1989). The average density may prove to be closer to 100 dwellings per hectare because there are many large infrastructure facilities in the area. Of the roughly 8,000 housing units, 80 per cent will be in four/six-storey blocks. The remaining 20 per cent will be divided equally between buildings of more than six storeys and two/four-storey units with gardens. Development costs were set at Dfl340 million (£139.3 million at exchange rates of March 1996) and it is estimated that Dfl170 million (£69.6 million at exchange rates of March 1996) will be returned to the public purse from land leased to developers and home-owners. The deficit will be carried by local and central governments, which indicates the extent to which this is a public project (Municipality of Amsterdam 1989).

In terms of urban design it is recognized that the area has a special character and urban qualities, but questions surrounding the strategy for urban design have strained relations between municipal and central governments, particularly when, in 1986, the latter subsidized a local community group to produce an alternative to the municipal government's proposals (Bongenaar and Veerkamp 1987). Ultimately, the work fell to external urban designers, and responsibility for the design of buildings is distributed between forty architects from several countries.

Initially, the date for completion of the project was 1996 (Municipality of Amsterdam 1991a). In 1995, however, less than half of the development was complete. The target date for completion has been reset for 1999, although the area is not likely to be fully integrated with the city until adjoining areas are developed.

There are potential problems, notably difficulties relating to access, transport and parking, which may result from relatively high densities. Nevertheless, the area's popularity is underpinned by good design and its historic character, and private housing commands relatively high prices for a socially mixed housing area.[3]

THE IJ-OEVERS AREA

From the beginning, the approach to the IJ-oevers or central part of the waterfront has been twofold: to strengthen the economic position of the inner city, and to incorporate the waterfront into the physical and functional structures of central Amsterdam. The area covers 330 ha, of which 172 ha is land (Fig. 11.3). As the principal landowner, the municipality was anxious to take control of planning, the provision of infrastructure and the distribution of land. This reflects the general pattern for public-sector developments in The Netherlands (Badcock 1994). In the period 1982–90, the planning process was dominated by discussions about building heights, the layout of connecting axes between the inner city and the IJ-boulevard, and whether there should be a two- or four-lane highway (possibly underground) behind the Central Station. The proportion of low-rent social housing that might be incorporated into the project was another factor in discussions. In addition, the feasibility of a large, high-quality shopping centre of 30,000 m^2 close to the Central Railway Station was examined. In planning circles it was generally agreed that retailing in Amsterdam's inner city could cope with the extension, but retailers feared the competition. The potential for office development was given relatively little exposure until 1991. Thus,

Figure 11.3 Amsterdam: IJ-oevers development model (1990)
Source: Amsterdam Waterfront Finance Company

many of the problems that later hindered the development of the IJ-oevers area were not exposed until the early 1990s.

It is important to note at this point that discussions about the IJ-oevers project had a political backdrop and were flanked by political events. Prior to 1990, the project was 'fronted' by the social democratic Partij van de Arbeid (PvdA) which actively favoured the development of the area. Their alderman, Michael Van der Vlis, took control of the project and, for example, defended it at public hearings and actively marketed it (see for example Bouw 1987). However, the 1990 elections brought a new local government and the PvdA handed over the position of alderman for urban planning to the more left-wing Groen Links party. The Groen Links alderman, Jeroen Saris, caught up in an increasingly ambitious project, was faced with the difficulty of explaining an 'alien' development to the Groen Links party. In addition, he had to operate without the backing of the PvdA, which stood to gain if Saris lost control of the project as well as his political support.

Saris' discomfort stemmed partly from the fact that, in 1991, the municipality produced a white paper that stressed the need to redevelop the IJ-oevers area for 'urban purposes' (Municipality of Amsterdam 1991b). In the same year, the AWF or Amsterdam Waterfront Finance Company (or Financieringsmaatschappij) was founded as a public–private partnership to shape a development centred on 'urban purposes'. This company was based on an equal partnership of the municipality and MBO/NMB, which is part of the large Dutch insurance and banking group ING. It seemed that Amsterdam was pushed into this partnership by central government. Moreover, grants for the development were linked to the progress of development proposals (*Volkskrant* 1990; *Parool* 1991).

The AWF was required to produce a business plan for the development of the IJ-oevers area. In March 1993, it produced a strategy that went further than the city's earlier white paper by increasing the overall density for the proposed development by some 25 per cent (AWF 1993). The ambitions of the AWF and the municipality raised the prospects for a major office centre on the IJ. Scheduled for development in 1995–2010, the AWF's proposals were significant for the economy of Amsterdam's inner city and the metropolitan region. It was also clear that, if developed, the proposals would have a significant impact on the physical structure of Amsterdam by extending the inner city, providing greater public access to the IJ and revitalizing a large strip of redundant land.

To develop the area, the AWF proposed that it should be subdivided in three companies to oversee:

(a) the preparation of the land and site development,
(b) the development of buildings, and
(c) the tasks associated with management and real estate.

It produced calculations to show that Dfl5.7–6.5 billion was a feasible budget for land acquisition, infrastructure and the development of the project. It argued that the development would generate 30,000 jobs over a fifteen-year period (Witbraad and Jorna 1993). However, ING (which made up the private-sector part of the AWF partnership) harboured doubts about the feasibility of the project. These misgivings were crucial, given the size of the investment which the project would require.

The AWF was subsequently dissolved and the proposed office element of the project foundered. After a year of silence, the municipality was anxious to confirm that the project had not been shelved (Municipality of Amsterdam 1994a); but the momentum behind the initial development proposals had evaporated, and the scale of the project has been reduced through re-zoning (Table 11.2).

The project may now proceed on the basis of a new strategy. In this respect, two aspects of the IJ-oevers project are instructive: the collapse of the AWF and the initial failure to develop office space in the area; the

Table 11.2 Amsterdam: distribution of land uses for the IJ project

	Municipality	*AWF*	*Municipality*
	1991	*1993*	*1994*
Land use (m^2)			
Commercial	517,200	668,600	—
Housing	344,400	424,000	478,000
Public facilities	93,600	151,600	—
Non-housing	—	—	461,000
Totals	955,200	1,244,200	939,000

Sources: Municipality of Amsterdam (1991b) (white paper); AWF 1993; Municipality of Amsterdam, zoning plan (1994b)

options for office development and the potential of the project to succeed under a new strategy. These are examined below.

THE FAILURE OF INITIAL PROPOSALS FOR THE IJ-OEVERS

A number of factors are cited for the failure to realize office development in the IJ-oevers area and the collapse of AWF as a public/private partnership. It is difficult to unravel or rank these, but they relate to the problems of: decreasing demand; the polynuclear nature of the Dutch and Amsterdam office market; the supply of office space in the Amsterdam region; the inflexibility of proposed plans for the IJ-oevers area; doubts about infrastructure grants; and, lastly, the sense of *incompatibilité d'humeur*, not to say distrust, between interests associated with the project.

Decreasing demand

The stress on the office component of the IJ-oevers project as its economic core meant that better car and rail access was needed. Equally, it was clear that improved transport links would be expensive. The project was enlarged to match rising cost estimates, but this made it more vulnerable to the common problems presented by demand, development cycles and market forces, and its capacity to compete with other office locations. It was also vulnerable to the difficulties of researching market forces and of making long-term estimates of the potential demand for office space. In 1989, the take-up of office space in the Amsterdam region was 450,000 m^2. From 1984 to 1993, however, the average level of annual demand was 270,000 m^2 (De Boer Den Hartog Hooft 1995). But demand is only part of the explanation for the failure to realize the initial office proposals for the IJ-oevers area.

Polynuclear office markets

Office space is located within the Randstad conurbation in several dispersed locations that show little differentiation in size, vacancy rates, rent levels, demand and the quality of office space (Jones Lang Wootton 1995; De Boer Den Hartog Hooft 1995; DTZ Zadelhoff 1994) (Fig. 11.4). On the

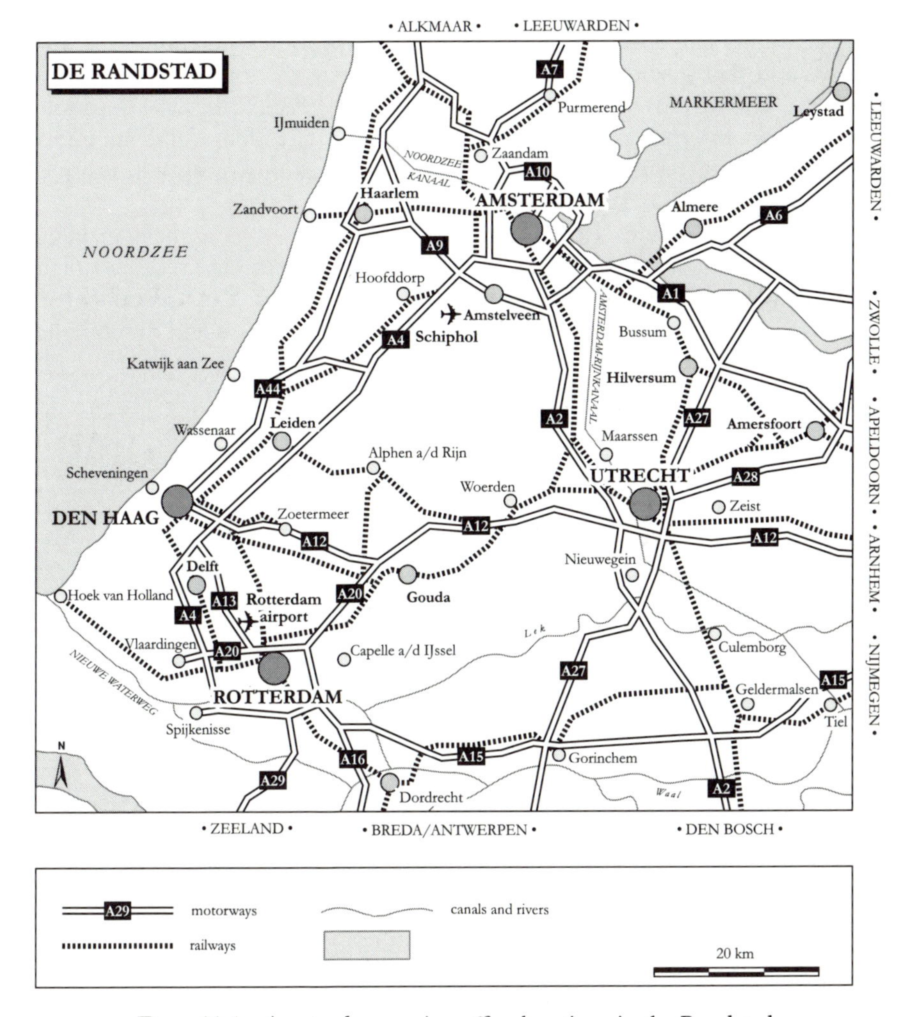

Figure 11.4 Amsterdam: major office locations in the Randstad
Source: Jones Lang Wootton 1993

whole, the Randstad conurbation is a high-quality, international economic centre. It is part of an advanced service economy, accommodates a wide variety of office functions and a large share of Holland's major multinational corporations and foreign multinationals. In the Dutch urban hierarchy, Amsterdam is the main financial centre and the major centre for international economic activities. As an 'office market', the Amsterdam municipality possessed approximately 4.5 million m^2 of office space in 1993. Rotterdam, as the centre for industry and transport, had 2.7 million m^2. The Hague, as the major political centre, had 3.2 million m^2 (Table 11.3).

Amsterdam does not, however, rank with Europe's larger office centres such as London or Paris (roughly 25 and 35 million m^2 of office space, respectively). The Randstad lacks the diversity of office functions, accessibility and other qualitative factors that characterize the top office centres in Europe. The majority of foreign companies in the Randstad serve domestic rather than international markets, although the city does fulfil an international role in terms of distributive and multinational corporations and financial services (Dwarkasing *et al.* 1988).

Bongenaar and Lie (1991) define top office locations as having: large homogeneous office areas; markets characterized by scarcity; relatively high rents; and greater development prospects. Scarcity is a consequence of demand, and of the limits placed on the size of office areas and on building heights. A top location has a critical mass (perhaps 2 million m^2) of advanced office functions, and serves a highly developed and diversified urban economy. There are no such office centres in The Netherlands. The larger Randstad office market, and the office markets of Amsterdam, Rotterdam, The Hague and Utrecht, have a secondary and polynuclear character (Korteweg and Lie 1992). In fact The Netherlands will never have an office location comparable with those of London or Paris and, in this respect, it can be argued that the goals of the AWF and the Ministry of Spatial Planning to realize a top international office location in The Netherlands were never feasible. Gradual change, from the present loose, polynuclear structure of office locations, to a clustered structure based on a small

Table 11.3 Randstad: office markets and average prime rents

	Office stock ('000,000 m^2)	*Prime rents (Dfl per annum)*
Amsterdam	4.5	450
The Hague	3.2	350
Rotterdam	2.7	285
Utrecht	1.5	325

Note: Stocks and rents are averaged over one/two years 1993/1994. Exchange rates (March 1996): Dfl 2.44 = £1; Dfl 1.57 = US$1

number of key centres is a more likely option. Under a 'key centre' policy, high-quality office space and major office users might be concentrated in one major location in each of the three or four major cities. However, this calls into question the issue of office location policy in the Randstad.

The prospects for office development on the waterfront can be examined in the context of the 'key centre' option. However, this focuses attention on the fact that the waterfront is in competition with several other office locations and, in particular, with the Amsterdam-South area. Thus, the potential of the IJ-oevers as a major regional centre for office development will depend on local government protection and a policy of positive discrimination in the development of office locations. This was underlined by INRO TNO as a consultant to the IJ-oevers project (INRO TNO 1993). The question is, however, how far can (or should) local government go in order to 'steer the market'.

Supply policy

Amsterdam's waterfront can yield an area for development that can draw benefits from the inner city and the IJ, and accommodate a mix of supportive functions within an attractive and historic environment. But the current popularity of peripheral and suburban office locations in Amsterdam and the Randstad poses a major problem for office development on the waterfront (Twijnstra Gudde 1994; Dewulf and de Jonge 1994). Office areas situated alongside motorways are particularly popular because they provide car access, parking and relatively low costs (Fig. 11.5). Conversely, the drawbacks of peripheral and motorway office locations could serve to highlight the advantages of the city centre and the waterfront. There are doubts about the environmental quality of peripheral, motorway office locations. They usually lack the support of shops, post offices, recreational facilities, restaurants and bars. They also lack the urban characteristics and social and cultural advantages of the inner city. There is no housing, and architectural and planning standards are generally poor.

In contrast, the waterfront could accommodate office space, housing, recreational and other supportive facilities, while providing some of the characteristics that make peripheral locations attractive to developers and office activities. The development of the IJ-oevers area was intended to exploit the opportunities of the site and its relationship to the inner city and to compete with peripheral office locations in terms of access and parking. Hence the importance given to improved car access (which, incidentally, poses a threat to the historic character of the inner city).

It is significant, however, that despite the advantages offered by the waterfront, ABN–AMRO chose to locate its new 70,000 m^2 headquarters in the Amsterdam-South area. Although rents in the area are relatively high, Amsterdam-South accommodates the World Trade Centre, the Insurance

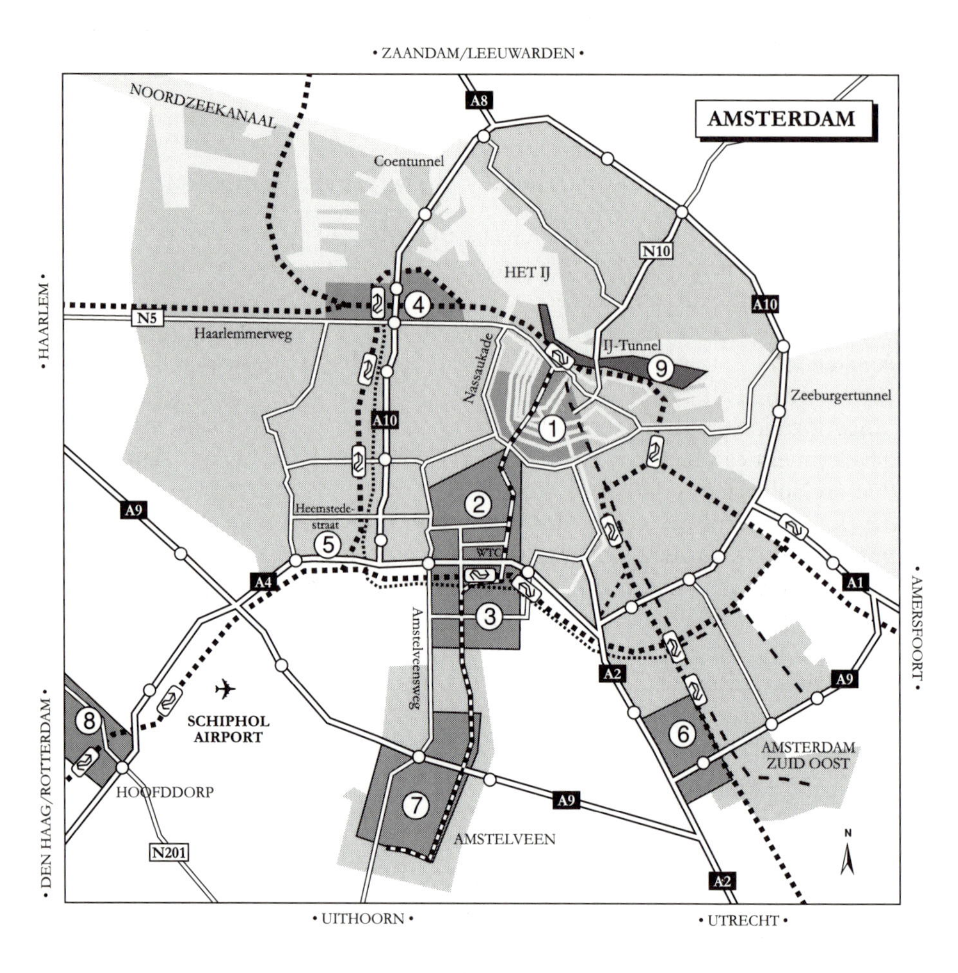

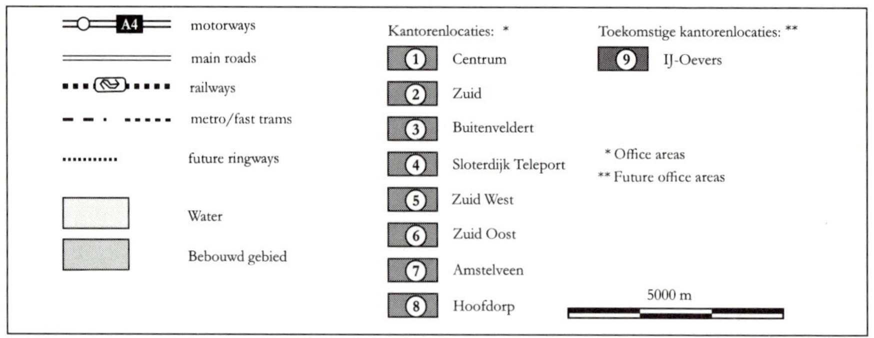

Figure 11.5 Amsterdam: major office locations in the Amsterdam area
Source: Jones Lang Wootton 1993

Exchange, the Cantony and County Court and several corporate headquarters. Presented with the option to locate on the waterfront in 1991–92, ABN–AMRO told the municipality that it was not willing to 'pioneer in a trench'.[4]

Central government policy and infrastructure grants

The regional context for office development has also created obstacles for the redevelopment of the IJ-oevers area. Thus, the prospects for office development on the waterfront changed in 1988, when national authorities declared that a network of high-quality office locations would be created in Amsterdam, Rotterdam, The Hague and Utrecht (Ministerie van VROM 1988, 1990). This policy is intended to address the fragmented pattern of office location in the Randstad (Table 11.3). Under this policy, an office development can be recognized as a 'key project' and win financial subsidies that are subject to conditions regarding the progress and execution of the development and the commitment of private investment. Projects are defined against a location policy for office employment, based on the nature and mobility of office users. The large headquarters of ministries and financial institutions (where employees have few external appointments and few visitors) are encouraged to relocate near inter-city railway stations in inner-city locations (labelled A locations). Companies that require accessibility by car and public transport are directed towards locations that provide those characteristics (B locations). Industries, transport and distribution companies are allowed to use motorway locations (or C locations) (Ministerie van VROM 1990b). However, this policy does not acknowledge the various costs and advantages of different locations and individual sites, or the fact that office users have different needs and preferences (Twijnstra Gudde 1994; Dewulf and De Jonge 1994). The office location policy in the Randstad ignores the fact that costs relating to land acquisition, treatment of polluted soil or sound-proofing vary between individual sites; and that the nature of supportive facilities and infrastructure differs between sites and different types of location. In effect, it does not consider the relative capacity of individual locations to generate rent and development profits and to compete for development capital. In this respect, it does not relate the flow of development capital into office development and official office location policies.

In terms of the waterfront in Amsterdam (which adjoins the Central Station) the Randstad's office location policy obscures the fact that developers and property interests do not necessarily value inner-city A-type locations over other locations. By linking the flow of government grants to the commitment of private investment, the government has created a dilemma for projects where the process of securing investment is tied to the security of grants and vice versa. This frustrates negotiations regarding investment, and places the municipality in a difficult position in terms of planning its expenditure on the waterfront.

Local commitment

The redevelopment of the IJ-oevers area has also been retarded by the attitudes of local authorities and public figures. The attitudes of government officials, councillors and aldermen differ between Amsterdam and Rotterdam (*Parool* 1990). Rotterdam is a city where things 'get done', perhaps because of its 'underdog position' and the absence of a historical inner city. Amsterdam, on the other hand, is a city where development proposals are debated. Politicians are mindful that community groups managed to block transport and infrastructure proposals for Amsterdam's inner city in the 1970s (de Hen 1986). There is evidence that the municipality wanted to expose the work of the AWF through public hearings, while the private interests involved wished to present a prepared strategy to the public (Alons and Partners 1994). This may also have indicated a lack of cohesion in the relationship between the public and private elements of the AWF. But whatever the causes, the political context for the redevelopment of the IJ-oevers area has been half-hearted. No major political figure has adopted a long-term pioneering stance on the project, thus depriving it of the strength of political support afforded to the Kop van Zuid development on the waterfront in Rotterdam (Teisman 1992).[5] The municipality of Rotterdam appears to be able to interweave the goals of citizens, developers, municipalities and state ministries. As a result a new bridge, subway station, 91,000 m^2 of office space (out of a planned 400,000 m^2) and the first 1,000 of a planned total of 5,300 dwellings have been built in the Kop van Zuid project. That these developments have appeared without recourse to public/private partnerships may be instructive in terms of the IJ-oevers project and the failure of the AWF.

The current situation

To compete effectively with other office locations, the IJ-oevers area could be enhanced by good infrastructure, greater accessibility by car and a high-quality, well-planned environment. The cost of these factors, however, can be met only through massive financial and administrative support from local and national governments. But any efforts made by national and local governments to force or cajole development capital on to the waterfront would fly in the face of official planning policy.

The prospect of preferential treatment raises an interesting dilemma for Dutch planning: namely, why should the waterfront be awarded public funding or special privileges? Why should the IJ-oevers area operate outside the control of regional policies that are supposedly aimed at equitable and good planning? Given that development capital is reluctant to flow into the IJ-oevers area, and that any state interference must be justified, the case for a special status for the waterfront can be made only on the basis of potential social gains and/or the benefits of correcting deficiencies in property markets. But, in reality, the massive financial incentives and political backing awarded to waterfront developments outside Holland are unlikely to be bestowed under Dutch planning.

If the waterfront is not to receive special treatment, the potential for office development on Amsterdam's waterfront depends on the 'natural capacities' of the IJ-oevers area as a large centrally located waterfront site. However, this argument leads back to the issue of the competition between office locations, and, ultimately, the waterfront may require a strategy that targets particular segments of the office market and recognizes the actual constraints and opportunities facing the area.

There is a wide range of office activities in the city centre, but the distribution of office employment and larger companies shows the predominance of suburban office locations (Table 11.1). While the inner city contains no more than 20 per cent of the total office stock in the Amsterdam region, the centre accommodates more relatively small firms. Firms with more than ten employees gravitate towards suburban locations. The motorway system and the general network of supporting services and links in the Randstad seems to favour suburban office location (Hessels 1992). In this respect, Amsterdam has surpassed the remarkable suburbanization of office space in cities in the United States (Fulton 1987; Pivo 1993; Garreau 1991). The degree of suburbanization is such that potential demand for office space in the inner city, or on the waterfront, is more likely to be uncovered by research into the pattern of location for different types of companies. For example, the inner city still has more insurance companies and banks, although the suburbs dominate in terms of the financial sector. Similarly, examination of the overall business sector shows that the tendency of different companies to locate outside the city centre varies according to the type of company. Most legal, accountancy and computer services are located outside the city centre, whereas engineering/architectural practices, advertising agencies, economic consultancies, news agencies, employment agencies and miscellaneous business services all prefer the city centre (Bongenaar 1995).

There are companies which are sensitive to the functional, 'environmental' and social qualities of the inner city. This lends support to the argument for a major office centre in the IJ-oevers area. Such a centre might draw off 5 per cent of the market for office space in the Randstad, and could stimulate new demand for office uses that are not well served by the relatively homogeneous Dutch office market. In theory, the waterfront could provide a common base for companies involved in 'culture', entertainment, design, fashion, marketing, advertising, publishing, broadcasting, recreation or tourism. Some of these could provide primary or secondary clusters or a critical mass of compatible occupiers. The project could be targeted directly at those interested in an 'urban base' and those who prefer to live and work in the inner city rather than in a standard motorway office location.

There are signs that the project may already be heading in this direction. The municipality has announced that it is examining the prospects for a range of subsidized organizations in the IJ-oevers area. Functions such as a theatre group, a centre for modern music and the central library might be housed in a 100,000 m^2 'cultural complex' (*NRC-Handelsblad* 1995). Furthermore, a new passenger terminal will be built to accommodate the roughly 100 cruise ships that pass through Amsterdam. In 1995 work began on the new 11,500 m^2 National Science Centre, close to the Central Station.

This strategy raises a number of questions. There is the issue of whether government (or government-funded) organizations can act as location leaders. It might be wise to attract, for example, a large broadcasting company into the IJ-oevers area, to demonstrate to investors that commercial development is feasible. While the potential for synergy between companies on the waterfront is important, it may be necessary to solicit demand from a wider range of potential occupiers.

CONCLUSIONS

Several lessons may be learnt from Amsterdam's waterfront. First, it is clear that the division of the area available for development between two projects allowed redevelopment in the eastern port area to proceed under its own momentum. The creation of the residential district is now well under way, and (at this stage) it is seen as a success. The IJ-oevers project, however, is in a different situation. Arguably, large and complex projects require a long-term perspective. Public and private interests need time to negotiate. A workable planning and development strategy takes time to prepare. Even in the embryonic stage, projects need to maintain momentum if public, political and private interests are to remain committed. For some years the IJ-oevers project was hot news. Currently, because there are virtually no tangible results, there is no media coverage.

The municipal government needs to clarify its intentions and to produce a clear organizational strategy for the IJ-oevers area, preferably with the political backing of a major public figure. The Municipality of Amsterdam's White Paper *Anchors in the IJ* (1994) may be a starting point in this process. Other interests that need to show commitment are: the national government, which is partly responsible for 'key projects', infrastructure and regional development; the provincial authorities that are responsible for regional policy; other municipalities; and the private sector in the form of investment and development interests. Whether the project requires the formation of a new public/private partnership remains to be seen. If a new partnership is formed, public and private interests need to share common objectives and to work from a position of trust. In addition, the project needs to involve the general public and the resident population of Amsterdam's inner city. There is also a need for further research on the prospects for development and its potential effects. Given its long-term framework, the project needs a 'master plan' and a detailed programme for phased development. In short, good planning, research, political support and public participation are necessary to the success of a project of this size.

While these factors may be put in place, the Dutch economy is slowly recovering from recession. However, it could be argued that recession should not be allowed to overshadow proposals for the IJ-oevers area. The prospects for the regional economy and the nature of the office market suggest that, in the medium term, a large-scale, good-quality project should succeed. The combination of good car access, public transport, supporting facilities, proximity to the core and an 'urbane' waterfront setting could provide a sound basis for development, particularly if this is augmented by strong political leadership, greater support from the public and private sectors, and good planning and design. Aspects of the development process may also be crucial, notably the installation (at an early stage) of appropriate infrastructure. The project also needs to attain a critical mass, perhaps by concentrating similar or compatible office activities in a development with a specific identity. In this respect, the nature of the first occupiers could be important in laying the foundations for further demand.[6]

Even if the conditions established above are met, however, the waterfront will not reach the status of an international office centre. It is likely to attract only certain segments of the office market. However, this factor may be turned to its advantage and, with the correct strategy, the IJ-oevers area could house a good development that would benefit the waterfront and the city. The key lies in adapting development proposals to real opportunities and constraints. Thus, Amsterdam's waterfront may hold useful lessons for projects that were launched in the 1980s, and which must now find a solid basis for development in the 1990s.

NOTES

1. Facilities might include the central library, theatres, a cinema complex, etc.
2. Research points consistently to the factors that determine user demand in local and national office markets. Emphasis is given to accessibility by car, parking, access to an international airport, public transport and the availability of modern buildings (Twijnstra Gudde 1994).
3. Typically, private dwellings sell for between Dfl200,000 and 550,000.
4. The fact that ING (ABN–AMRO's main Dutch competitor) was the leading developer for the IJ-oevers area may have contributed to their rejection of the waterfront.

5. So far the Kop van Zuid has tended to attract mainly government offices. As government offices generally are not seen as prime tenants or occupiers in The Netherlands, this qualifies the success of the project as an office location.
6. In this respect, government offices are not usually held in high esteem by the Dutch business community and should not be allowed to front the occupation of the waterfront. In The Netherlands, the ideal 'location leaders' are the headquarters of large and preferably multinational corporations. However, these qualifications are overshadowed by the more fundamental question of demand for office space on the waterfront.

BIBLIOGRAPHY

Alons and Partners (1994) *Evaluatie rapport IJ-oevers* (Evaluation report IJ-oevers), The Hague: Alons.

Amsterdam Waterfront Financieringsmaatschappij (AWF) (1993) *Ondernemingsplan ontwikkeling IJ-oevers Amsterdam* (Business Plan, Development IJ-oevers Amsterdam), Amsterdam: AWF.

Badcock, B. (1994) 'The strategic implications for the Randstad of the Dutch property system', *Urban Studies* 31: 425–45.

Bongenaar, A. (1995) *Vestigingsplaats-factoren en ruimtelijke Structuur* (Location factors and urban and regional form), Delft: INRO–TNO.

Bongenaar, A. and Lie, R.T. (1991) 'Naar hoogwaardige kantoorlocaties' (Towards high-level office locations), *ESB* 7/8: 792–6.

Bongenaar, A. and Veerkamp, J. (1987) 'Stad aan het IJ' (City along the IJ), *Intermediair* 23(14): 27–31.

Bouw (1987) 'Bouwen in havens van Amsterdam en Rotterdam' (Building in the ports of Amsterdam and Rotterdam), *Bouw* 8: 17–14.

De Boer Den Hartog Hooft (1995) *Kantorenmarkt Amsterdam 1994* (The Amsterdam office market 1994), Amsterdam: De Boer Den Hartog Hooft.

Dewulf, G. and de Jonge, H. (1994) *Toekomst van de kantorenmarkt 1994–2015* (Future of the office market 1994–2015), Delft: TU.

DTZ Zadelhoff (1994) *Visie achter de feiten* (Vision behind facts), Utrecht: DTZ/Zadelhoff.

Dwarkasing, W. D., Hanemaayer, D., van der Mark, R. and de Smidt, M. (1988) 'Ruimte voor hoogwaardige kantoren' (Space for high-level offices), *Nederlandse Geografische Studies*, 71, Utrecht: Universiteit van Utrecht.

Fulton, W. (1987) 'Offices in the Dell', *Planning* 52/7: 13–17.

Garreau, J. (1991) *Edge City, Life on the New Frontier*, New York: Doubleday.

Hart, Ter, H. W. (1984) 'Tussen twaalf en twee' (Between twelve and two), *ESB* 4/1: 8–9.

Heinemeijer, W. F., Wagenaar, M. F. *et al.* (1987) *Amsterdam in kaarten* (Amsterdam in maps), Ede (The Netherlands): Zomer and Keuning Boeken B.V.

Hen, de P. E. *et al.* (1986) *Om het behoud van een explosieve stad* (For the conservation of an explosive city), Muiderberg: Coutinho.

Hessels, M. (1992) *Locational Dynamics of Business Services: An Intrametropolitan Study on the Randstad Holland*, Utrecht/Amsterdam: Royal Geographical Society/Faculty of Geographical Sciences, Utrecht University.

Hoyle, B. S., Pinder, D. A. and Husain M. S. (eds) (1988) *Revitalising the Waterfront: International Dimensions of Dockland Redevelopment*, London: Belhaven Press.

INRO TNO (1993) *Kantoorfunctie voor de IJ-oever* (Office functions for the IJ-oevers), Delft: INRO TNO.

Korteweg, P. and Lie, R. (1992) 'Prime office locations in the Netherlands', *Tijdschrift voor Economische en Sociale Geografie* (TESG) 83(4): 250–62.
Jones Lang Wootton (1993) *City Reports: Amsterdam, The Hague, Rotterdam, Utrecht*, Amsterdam: Jones Lang Wootton.
—— (1995) *The Dutch Office Market*, Amsterdam: Jones Lang Wootton.
Ministerie van Volkshuisvesting, Ruimtelijke Ordening en Milieubeheer (1988) *Vierde nota over de ruimtelijke ordening extra* (Fourth White Paper on spatial planning), The Hague: Ministerie van VROM.
—— (1990a) *Vierde nota over de ruimtelijke ordening extra*, Den Haag: Ministerie van VROM.
—— (1990b) *Werkdocument geleiding van de mobiliteit door een locatie beleid voor bedrijven en voorzieningen*, The Hague: Ministerie van VROM.
Municipality of Amsterdam (1980–90) *Amsterdam in Cijfers* (Amsterdam in figures), Amsterdam: Municipality of Amsterdam.
—— (1985a) *Nota van uitgangspunten Oostelijk Havengebied* (White Paper on eastern port area), Amsterdam: Municipality of Amsterdam.
—— (1985b) *Structuurschets Oostelijk Havengebied* (Structure vision Eastern Port Area), Amsterdam: Municipality of Amsterdam.
—— (1986) *IJ-oevers en Oosterdok, verkenningen en plannen* (IJ-oevers and Eastern Dock: explorations and plans), Amsterdam: Municipality of Amsterdam.
—— (1989) *Nota van uilgangspuntern Oosterlijk Havengbied* (White Paper on eastern port area), Amsterdam: Municipality of Amsterdam.
—— (1991a) *KNSM-Stadseiland: de nieuwe thuishaven van Amsterdam* (KNSM City island: the new home port of Amsterdam), Amsterdam: Projectgroep Oostelijk Havengebied.
—— (1991b) *Nota van uitgangspunten IJ-oevers* (White Paper on IJ-oevers), Amsterdam: Municipality of Amsterdam.
—— (1994a) *Ankers in het IJ* (Anchors in the IJ), Amsterdam: Municipality of Amsterdam.
—— (1994b) *Bestemmingsplan zuidelijke IJ-oevers* (Zoning plan: southern IJ-oevers), Amsterdam: Municipality of Amsterdam.
—— (1995) *De Amsterdamse kantorenmarkt in 1994* (The Amsterdam office market in 1994), Amsterdam: Municipality of Amsterdam.
NRC-Handelsblad (1995) 'Amsterdam wil cultuurcomplex aan Y-oevers' (Amsterdam wants a culture complex along the Y-oevers), *NRC-Handelsblad* (6 June).
Parool (Het) (1990), 'IJ-oevers: unieke kans of megalomanie' (IJ-oevers: a unique chance or megalomania: interview with the former Rotterdam alderman Jan van der Laan), *Het Parool* (26 May).
—— (1991) *IJ-oevers Special* (6 June).
Pivo, G. (1993) 'A taxonomy of suburban office clusters: the case of Toronto', *Urban Studies* 30(1): 31–49.
Teisman, G. R. (1992) *Complexe besluit-vorming: een pluricentrisch perspectief op besluitvorming* (Complex decision-making: a pluricentric perspective on decision-making), Gravenhage: VUGA.
Twijnstra Gudde (1994) *Nationaal kantorenmarktonderzoek 1993*, Amersfoort: Twijnstra Gudde.
Volkskrant (De) (1990) 'Rijk eist voortvarendheid bij plannen IJ-oever' (Central government demanding energy in IJ-oevers development), *De Volkskrant* (15 November).
Witbraad, F. and Jorna, P. (1993) 'Waterfront regeneration: the Y embankment project in Amsterdam', in J. Berry, S. McGreal and B. Deddis (eds) *Urban Regeneration: Property Investment and Development*, London: Spon.

12

CONCLUSIONS

Patrick Malone

INTRODUCTION

The waterfront projects examined in previous chapters illustrate a wide range of issues. Some are specific to individual projects, others are common to the international framework for urban development. However, each project has been generated by a set of underlying forces that raise questions about the nature and power of property and political interests in urban development, and the weight given to planning and social objectives. London Docklands demonstrates the power of political and property interests to restructure planning to fit the 'enterprise culture', and the conversion of a planning system to strategies based on the market and short-term opportunism (Thornley 1991; Fainstein 1994). Docklands, and particularly the Enterprise Zone, demonstrated the use of a wide range of planning mechanisms that favoured the developer. The minimization of development controls, simplified planning processes and the state-sponsorship of the private property sector were portrayed as positive means of promoting urban development. However, the projects covered in this book illustrate that the exploitation of development processes by political and property interests takes place within different local contexts and planning frameworks.

In the 1980s, the context for the development of Dublin's Custom House Docks was akin to that of London Docklands. Both projects were managed by special development agencies and both were heavily supported by the state through financial incentives. In the Custom House Docks, however, property and national economic ambitions were united in the development of a new financial node in the global economy, created to draw finance capital into Ireland. The financial incentives used to support physical and economic development were effectively pooled.

Sydney's Darling Harbour project also demonstrated the remarkable degree of political support awarded to waterfront projects in the 1980s, although the political and economic forces behind it differed from those that reshaped the waterfronts of London and Dublin. All three projects illustrated that the waterfront could provide a 'shop-front' in which cities and states could set out the visual symbols of the 'new' economy. However, much of the support for Darling Harbour stemmed from its political value at a local level, the ambitions of Premier Neville Wran and Australia's bicentennial celebrations.

In Sydney, politicians paid a price for their investment in the vote-catching potential of the waterfront. But Neville Wran was not alone in using urban development as a political stage during the period of the *Grands Projets* in Paris, the IBA programme in Berlin and the Olympics-based projects in Barcelona (Sudjic 1993). Factors such as economic deregulation, the concept of the 'world city' and the rise of urban marketing helped to politicize the waterfront. Through heavy urban marketing, projects could be portrayed in terms of urban renewal, housing, social, cultural and recreational facilities. Images of attractive urban developments could be combined with the prospects for increased employment and potential benefits to urban, regional or national economies.

In Japan, large waterfront and marine projects were also given strong political support in the 1980s. That support, however, was channelled through established planning mechanisms and existing rather than new public agencies. Given the 'applied' nature of Japanese planning, large marine developments have a two-way economic function. They are developed under the umbrella of metropolitan planning as extensions of the city and the urban economy, and they bolster the economy at the international level.

Clearly, differences in the context of waterfront development are significant if, for example, conditions in Britain are compared with those of Japan. Moreover, the situations for waterfront development in both countries contrast, for example, with that of The Netherlands where local authorities have a relatively large (and arguably a more social) role in the development process (Kreukels 1992: 242). The Dutch ambivalence about the pursuit of 'world city' status for Amsterdam has meant that office proposals for the IJ waterfront have had less backing than, for example, those in Dublin's Custom House Docks. Similarly, Toronto's ambitions for an ecosystem approach to planning contrast with that of the British 'hard right', which held that the LDDC was not 'a welfare association but a property based organisation' (Nigel Broakes, quoted in Ambrose 1986: 228).

While generalizations about the effects of the international economy or the political functions of urban development contribute to an overall view of waterfront projects, individual projects reveal local differences in development frameworks and in approaches to planning tools and social values. The differences in the economic and political forces behind developments are reflected in the physical and functional structures of individual projects. For example, the origins of London Docklands are expressed in incoherent spatial and land-use patterns and in the failure to develop good transport links with the rest of London (Edwards 1993). The LDDC's position on planning was also evident in its haphazard approach to property markets and development economics (Budd and Whimster 1992: 23). But again, Docklands contrasts with other waterfront projects, for example in Sydney, Barcelona and Genoa – projects which are seen to be relatively successful in physical and social terms (*The Architectural Review* 1989: 39–77).

In summary it may be said that waterfront developments are generated by common and individual factors. They reflect different local contexts for planning and factors that can be traced back to the international framework for urban development, but in so far as they have common origins they raise general questions about the future for waterfront development. Perhaps the most obvious question follows from the fact that many waterfront developments were launched by political and economic interests that were characteristic of the 1980s. Given that the factors which drove urban development in the 1980s are now diminished or spent, questions arise as to the nature of the forces that will carry waterfront development into the next century.

THE FUTURE FOR THE WATERFRONT

As the projects examined above demonstrate, ambitions invested in the waterfront in the 1980s have tended to collapse in the 1990s. Projects that were aimed at the global economy are proving to be out of scale with failing local property markets. Developments centred on the demand for office space can no longer rely on economic expansion or growth in the financial sector. It is significant, for example, that an estimated 20,000 jobs were lost in the financial services sector in the City of London over the five years from 1987 to 1991 (*Estates Times* 1991).

The internationalization of the waterfront has meant that projects are susceptible to world markets and a common economic recession. While London created a large display of financial muscle in Docklands, the bankruptcy of Canary Wharf in 1992 exposed the sensitivity of waterfront developments to international economic factors. Docklands also illustrated that the level of official support for projects exacerbated the sense of failure that enveloped many waterfronts in the late 1980s. Ultimately, the 'hype' surrounding projects reinforced an image of investment-led and politically contrived development.

The economic costs of development failures are carried by the public, which must pay the price for institutional lending practices and government sponsorship of private development. The huge outstanding loans of Olympia and York, and the great weight of property-based debt in the United States and Britain, are now part of the backdrop for waterfront development in the 1990s (Merrifield 1993: 1247–59). One consolation for financial institutions and governments may be that waterfront office space and property investments may be absorbed at rents and building prices substantially lower than in the late 1980s (Daniels and Bobe 1993: 551). Governments may also find comfort where part of the cost of failing waterfront projects is borne by banks and financial institutions in other countries. The failure of Olympia and York illustrated the international spread of bank and institutional lending behind the Reichmann empire (Fainstein 1994: 216). Similarly, the internationalization of capital may hold some consolation for the British banking system, in that this reduced the share of domestic banking capital in the UK property sector, from roughly 80 to 59 per cent in 1980–89 (Debenham, Tewson and Chinnocks 1989: 8). However, while international capital flooded into the UK property sector in the 1980s, particularly from Japan, British investors operating outside Britain were also exposed to the dangers of foreign property investment (Debenham, Tewson and Chinnocks 1990: 2). It should also be noted that, in Britain and elsewhere, recession has also tended to increase the financial costs of waterfront developments.

The cost of failing projects must also be counted in political terms. The level of political investment in the waterfront in the 1980s raises issues for the 1990s. In the face of economic recession, emerging political regimes have abandoned or denounced faltering developments initiated by outgoing or defeated politicians. The abandonment by Tokyo's new governor of the proposed World City Expo in 1995 provides a good example (Chapter eight). The lesson for urban development must be that projects that are hoisted on to the political platform by one regime are likely to be neglected, abandoned or thrown out by succeeding or opposing political regimes – particularly when projects falter economically or require massive support from the public purse.

The future for waterfront developments that are no longer driven by the political and economic forces of the 1980s hinges on finding new economic and political foundations in the 1990s. The effects of recession and the removal of political support can be seen in a number of retarded, shelved and abandoned projects. Some, like Antwerp or Amsterdam, were late starters in terms of national and international economic and development cycles. However, many projects have faltered in the light of changes in local and global markets, political frameworks and ruling ideologies. In some projects, the state may be drawn in to make good the losses in development profits, to provide infrastructure (which the public sector may have failed to provide), or to take a larger role in the development process. Projects may require new forms of partnership or the re-alignment of public and private interests (Brownill 1990: 152–70). But given the distinction between the agencies or frameworks for development and the economics of the development process, ultimately, developments must pursue real economic possibilities and work within the market constraints of the 1990s.

This may affect the physical nature of developments in several ways.

Where projects centred on office space are vulnerable to existing market conditions, developers may choose to target leisure, housing or other markets. As in the case of Amsterdam, office projects may also be restructured to exploit local demand and limited market opportunities. But these solutions can upset initial estimates of development profit, and may increase dependence on state funding and the need to provide infrastructure from public funds.

Together, the potential solutions for faltering developments, and the prospect of greater public-sector involvement, raise the issue of the future for planning. Changes in the nature of the forces that control development, and the stress placed on consensus and partnership, prompt the question 'are we back to planning?' (Brownill 1990: 152). However, this leads to further questions about the state of the planning and design professions, the culpability of planners and designers as regards the quality of existing developments, and the capacity of professionals to forge better strategies for the future.

While the sample of projects examined in this book is too small to justify detailed criticism, it suggests that waterfront development in the 1980s cannot be said to represent a high point in planning and urban design. Architecturally, the 'waterfront era' can be characterized by the spread of international 'architectural capital'. But the internationalization of practice has gone hand-in-hand with hybrid and imitative architecture based on the aesthetics of power and the rental value of style (Harvey 1989b: Zukin 1992). There is a feeling that development strategies lack depth and display an inability to generate comprehensive formulae for planning and design (Ambrose 1994: 177–86; *The Architectural Review* 1987: 31–7). Oversized and supply-led development programmes have produced projects and development schedules that are disproportionate to levels of demand. The LDDC's programme has been denounced as 'a very silly brief', which betrayed its ignorance of urban processes (Ambrose 1994: 186). But the inadequacies of 'silly' development programmes are compounded where they are overlain with weak and imitative design strategies. International design clichés now echo between the waterfronts of Boston, Tokyo and Dublin, as architecture is pressed into service as a marketing tool. Overtly form-based approaches to design provide weak criteria for land-use planning and frustrate the evolution of successful functional structures. These weaknesses take on a greater significance where urban design rather than planning is employed to generate frameworks for development (which research might show to have been a feature of the 1980s).

While there are doubts about the role of planning and design, their relationship to marketing and their capacity to provide strategies for the future, some good planning and architecture has emerged on the waterfront – even in London. Looking ahead, fresh opportunities for waterfront development may lie in the realm of housing, where planning can contribute to new policies for the reinforcement of inner-city populations, transport and the 'sustainable city'. As Greenberg suggests, housing projects may be at the forefront in terms of the return to social values in urban space, and the integration of new areas into existing spatial structures (Chapter nine). The improvement of public access to the waterfront and the provision of better public facilities that cater for wider social needs, also constitute challenges for the future. In this respect, the design of social space needs to get further away from the reproduction of 'ersatz urbanism' and crass historicism. A wider view has been taken by critics of the Subcentre project in Tokyo Bay, who put forward alternative development strategies based on the opportunities for wildlife areas and public open space: strategies that suggest a more profound approach to public amenities and social values (Chapter eight). As the redevelopment of Amsterdam's waterfront illustrates, the design and marketing of office space is another area of opportunity for planning systems that can evolve appropriate development controls and marketing strategies.

Waterfront development is also confronted by a wide range of issues concerning the economic, spatial and transport linkages between large waterside developments and parent cities. The projects covered in this book suggest that many developments are not well connected to parent cities in economic, spatial or functional terms. As Goldrick and Merrens point out, large waterfront developments also pose opportunities for the evolution of comprehensive planning policies that incorporate wider ecological perspectives (Chapter ten).

The challenges that face waterfront development in the 1990s should be seen as opportunities for good planning and design. However, the projects covered in this book demonstrate the power of economic and political forces in urban development and how these constrain planners and designers. But these projects also expose weaknesses in planning and design. Thus, the future for the waterfront hinges not only on the changing nature of political and economic forces in urban development, but also on the capacity of planners and designers to rise to the opportunities of the 1990s.

BIBLIOGRAPHY

Ambrose, P. (1986) *Whatever Happened to Planning?*, London: Methuen.
—— (1994) *Urban Process and Power*, London: Routledge.
The Architectural Review (1989) 'Dockland development', *The Architectural Review* 1106 (April): 27–88.
Ashworth, G. J. and Voogd, H. (1990) *Selling the City: Marketing Approaches in Public Sector Urban Planning*, London: Belhaven.
Brownill, S. (1990) *Developing London's Docklands: Another Great Planning Disaster*, London: Paul Chapman.
Buchanan, P. (1989) 'Quays to design', *The Architectural Review* 1106: 39–44.
Budd, L. and Whimster, S. (eds) (1992) *Global Finance and Urban Living: A Study of Metropolitan Change*, London: Routledge.
Daniels, P. W. and Bobe, J. M. (1993) 'Extending the boundary of the City of London? The development of Canary Wharf', *Environment and Planning A* 25: 539–52.
Davies, C. (1987) 'Ad hoc in the docks', *The Architectural Review* 1080: 31–7.
Deakin, N. and Edwards, J. (1993) *The Enterprise Culture and the Inner City*, London: Routledge.
Debenham, Tewson and Chinnocks (1989) *Money into Property*, London: Debenham Tewson Research.
—— (1990) *Special Rreport: Overseas Property Investment in the UK Property Market*, London: Debenham Tewson Research.
Edwards, B. (1993) 'Deconstructing the city: the experience of London Docklands', *The Planner* (February): 16–18.
Estates Times (1991) 'City of London: an Estates Times survey' *Estates Times* 1084: 23.
Fainstein, S. S. (1991) 'Promoting economic development: urban planning in the United States and Great Britain', *Journal of the American Planning Association* 57(1): 22–33.
—— (1994) *The City Builders*, Oxford: Blackwell.
Harvey, D. (1989a) 'Downtowns', *Marxism Today* 33(1): 21.
—— (1989b) *The Condition of Postmodernity*, London: Blackwell.
Kearns, G. and Philo, C. (1993) *Selling Places: The City of Cultural Capital, Past and Present*, Oxford: Pergamon.
Kreukels, A. (1992) 'The restructuring and growth of Randstad cities', in F. M. Dieleman and S. Musterd (eds) *The Randstad: A Policy Laboratory*, Dordrecht: Kluwer Academic Publishers.
Merrifield, A. (1993) 'The Canary Wharf debacle', *Environment and Planning A* 25: 1247–65.

Sudjic, D. (1993) *The 100 Mile City*, London: Flamingo.
Thornley, A. (1991) *Urban Planning Under Thatcherism: The Challenge of the Market*, London: Routledge.
Zukin, S. (1992) 'The city as a landscape of power: London and New York as global financial capitals', in L. Budd and S. Whimster (eds) *Global Finance and Urban Living: A Study of Metropolitan Change*, London: Routledge.

INDEX

Figures and tables are in *Italic*